Advanced Technologies and Societal Change

This series covers monographs, both authored and edited, conference proceedings and novel engineering literature related to technology enabled solutions in the area of Humanitarian and Philanthropic empowerment. The series includes sustainable humanitarian research outcomes, engineering innovations, material related to sustainable and lasting impact on health related challenges, technology enabled solutions to fight disasters, improve quality of life and underserved community solutions broadly. Impactful solutions fit to be scaled, research socially fit to be adopted and focused communities with rehabilitation related technological outcomes get a place in this series. The series also publishes proceedings from reputed engineering and technology conferences related to solar, water, electricity, green energy, social technological implications and agricultural solutions apart from humanitarian technology and human centric community based solutions.

Major areas of submission/contribution into this series include, but not limited to: Humanitarian solutions enabled by green technologies, medical technology, photonics technology, artificial intelligence and machine learning approaches, IOT based solutions, smart manufacturing solutions, smart industrial electronics, smart hospitals, robotics enabled engineering solutions, spectroscopy based solutions and sensor technology, smart villages, smart agriculture, any other technology fulfilling Humanitarian cause and low cost solutions to improve quality of life.

Ch Satyanarayana · Xiao-Zhi Gao ·
Choo-Yee Ting · Naresh Babu Muppalaneni
Editors

Proceedings of the International Conference on Computer Vision, High Performance Computing, Smart Devices and Networks

CHSN-2020

Springer

Editors
Ch Satyanarayana
Department of CSE
Jawaharlal Nehru Technological University
Kakinanda, Andhra Pradesh, India

Xiao-Zhi Gao
School of Computing
University of Eastern Finland
Kuopio, Finland

Choo-Yee Ting
Faculty of Computing and Informatics
Multimedia University
Cyberjaya, Malaysia

Naresh Babu Muppalaneni
Department of Computer Science
and Engineering
National Institute of Technology Silchar
Silchar, Assam, India

ISSN 2191-6853 ISSN 2191-6861 (electronic)
Advanced Technologies and Societal Change
ISBN 978-981-19-4043-9 ISBN 978-981-19-4044-6 (eBook)
https://doi.org/10.1007/978-981-19-4044-6

This Springer imprint is published by the registered company Springer Nature Singapore Pte Ltd.
The registered company address is: 152 Beach Road, #21-01/04 Gateway East, Singapore 189721, Singapore

Committee

Chief Patron

Prof. M. Ramalinga Raju, Hon'ble Vice-Chancellor, JNTUK, Kakinada

Patrons

Prof. G. V. R. Prasada Raju, Rector, JNTUK, Kakinada
Prof. B. Bala Krishna, Principal, UCEK, JNTUK, Kakinada

Honorary Chair

Prof. Allam Appa Rao, Ph.D(Engg).,D.Sc(h.c)., Former Vice-Chancellor, JNTUK, Kakinada; Chairman, IBCB, Visakhapatnam

General Chairs

Prof. Ch Satyanarayana, Registrar, JNTUK, Kakinada
Prof. S. V. Rao, Head-CSE, IIT Guwahati
Prof. Xiao-Zhi Gao, School of Computing, University of Eastern Finland, Finland

Program Chairs

Dr. Ch Sudhakar, National Institute of Technology, Warangal, Telangana
Dr. Naresh Babu Muppalaneni, National Institute of Technology, Silchar, Assam

Convenor

Dr. D. Haritha, Professor and Head in CSE, JNTUK, Kakinada

Organizing Secretary

Dr. B. N. Jagadesh, Professor and Head in CSE, Srinivasa Institute of Engineering and Technology, Amalapuram

Preface

The International Conference on Computer Vision, High Performance Computing, Smart Devices and Networks (CHSN-2020), is aimed to bring researchers together working in this area to share their knowledge and experience. In this conference, topics of contemporary interests would be discussed to provide a holistic vision on the latest technologies in computer science and engineering. The scope includes data science, machine learning, computer vision, deep learning, artificial intelligence, artificial neural networks, mobile applications development and Internet of Things, etc. Conference participants are expected to gain relevant knowledge and better understanding of the applications of computer science in various fields.

CHSN-2020 would be both stimulating and informative with the active participation of a galaxy of keynote speakers. We would like to thank all the authors who submitted the papers, because of which the conference became a story of success. We also would like to express our gratitude to the reviewers, for their contributions to enhance the quality of the papers. We are very grateful to the keynote speakers, session chairs and committee members who selflessly contributed to the success of CHSN-2020. We are very thankful to Jawaharlal Nehru Technological University, Kakinada, for providing the basic requirements to host the CHSN-2020.

Kakinanda, India
Kuopio, Finland
Cyberjaya, Malaysia
Silchar, India

Ch Satyanarayana
Xiao-Zhi Gao
Choo-Yee Ting
Naresh Babu Muppalaneni
Editors, CHSN-2020

Advisory Board

Contents

Chapter 1
Character Recognition System Using CNN for Sanskrit Text

R. Dinesh Kumar, M. Kalimuthu, and B. Jayaram

Introduction

Documents are in the form of papers that are understood by the human, but it is not possible for the computer system. So these documents are to be converted into a computer system readable form. OCR is the procedure of converting the scanned document into handwritten or printed text, symbols, letters and numerals and also can convert it into a possible format, for example, ASCII code [1]. Typically OCR can be used for processing the handwritten character and pattern recognition and is motivated greatly by a desire to enhance machine-to-man communication. Presently, few products are available for handwritten character recognition process, though different kinds of methods and procedures have been proposed [2].

Sanskrit Language

Sanskrit language is no longer spoken but still contains written text.

There are a few beliefs about the name 'Sanskrit':

R. Dinesh Kumar (✉)
Department of Computer Science and Engineering, Siddhartha Institute of Technology and Science, Hyderabad, India
e-mail: me.dineshkumar@gmail.com

M. Kalimuthu
Department of Computer Science and Engineering, Malla Reddy Institute of Technolgy, Hyderabad, India

B. Jayaram
Department of Computer Science and Engineering, RMK Engineering College, Chennai, Tamilnadu, India

Ch. Satyanarayana et al. (eds.), *Proceedings of the International Conference on Computer Vision, High Performance Computing, Smart Devices and Networks*, Advanced Technologies and Societal Change,
https://doi.org/10.1007/978-981-19-4044-6_1

1. Sanskrit is termed as the voice of Devas.
2. Sanskrit is also termed as Devanagari because of its excessive utilization in Brahmins of Gujarat.
3. An additional viewpoint is in that Devnagar area of Kashi, therefore it was termed as Devanagari (Sanskrit).

Sanskrit is the most precise scientific basis language. For a long period, it has been Indo Aryan's script language. It is also utilized by Marathi, Hindi and Nepali languages. Widely spoken language is Hindi since its script is in Sanskrit and Sanskrit has got the dialect status. In the initial stage, Hindi was stated as the state language and Sanskrit as the start script of the few states like Haryana, Madhya Pradesh, Himachal, Uttaranchal, Bihar and so on [3]. Currently, it is found that Sanskrit is connected with every other script. In this script, all letters are equal which means there is no concept of small or capital letters and is half syllabic in nature [4, 5].

Problems in Sanskrit Text

- All separate characters are fused by a headline named 'ShiroRekha' in the case of Sanskrit script. This fusion system creates it hard to isolate separate characters from the single words.
- There are different kinds of isolated dots, which are vowel modifiers, for example, 'Chandra Bindu', 'Visarga', and 'Anuswar', which add up to the confusion.
- Descender and Ascenders recognition are also difficult, attributed to the difficult nature of language.
- It comprises composite characters.
- Minor differences in same characters.
- It comprises a huge volume of stroke and character classes.

Vowels and Consonants

Sanskrit script comprises 18 vowels ('svar') and 34 consonants ('vyanjan'). In addition, vowels and consonants are also comprised of vowel modifiers named matra's which are located at right or left part of the Sanskrit script.

Literature Survey

The framework of Sanskrit Character Recognition system is shown in Fig. 1.1.

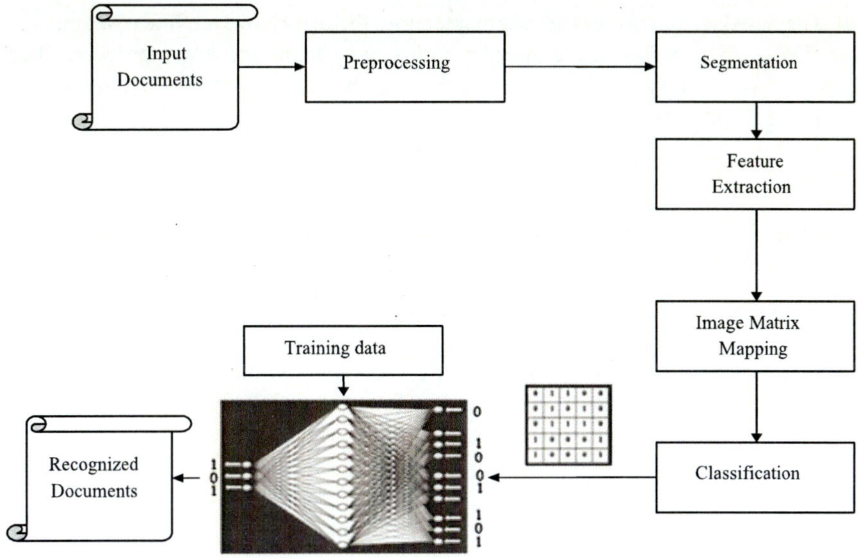

Fig. 1.1 Character recognition system

Input Documents

The documents are scanned with the proper scanner to provide the digital image for recognition. After this process, input image size is specified by the user which means the length and width of the document [6].

Preprocessing

After scanning, the next step is to remove noise from the scanned image [7]. The noise-free image is checked for skewness. Skewness is defined as the tilt (angle) in the bitmapped image of the scanned document. It is normally caused if the document is not straightly inserted into the scanner. But, most of the character recognition procedure is sensitive to the skew (orientation) of the input document, cropping is essential to implement the algorithms which can correct and detect the skew automatically.

Segmentation

After preprocessing, the noise-free image is passed to the segmentation phase, where the image is decomposed into individual characters. Sanskrit text character recognition process is carried out by applying the line and character segmentation techniques

[8]. The Sanskrit text images are segmented into the lines and each line is segmented into the words in terms of the upper modifier, consonant and lower process. Then the segmented words are converted into the straight lines which are used in Sanskrit text recognition process.

Feature Extraction

The feature extraction process examines a character segment and chooses a set of features that can be utilized uniquely to recognize the segmented text or character. The selection of a representative and stable set of a feature is the heart of the character recognition system. Different kinds of features are extracted to perform classification [9].

Classification

Classification is an important stage that is used for main decision-making process which is done by using the extracted features from the previous stages. The classification process identifies the character according to their preset rules [10]. The classification process is making decisions by using the class membership pattern. But this task is difficult because of the decision rule. Thus the feature extraction scheme is applied for reducing the misclassification probability. After this process, the classification process is done, but classification becomes an issue where characters fall into an unknown pattern.

Naïve Bayesian Classifiers

The Bayesian classifier is known to be capable of universal approximation and the output of a Bayesian network can be related to Bayesian properties. The Bayesian network has three input layers, namely input layer, hidden layer and the output layer in which each layer consumes the non-linear inputs and produces the linear output.

SVM Classifiers

The optimized features are applied to the Support Vector Machine which chooses the maximum fitness value to recognize the handwritten characters with better results. The extracted zones are considered as the features and are classified by applying the Support Vector Machine.

In this type, error function is minimized.

$$\frac{1}{2}w^r w + C\sum_{i=1}^{N} \xi_i \tag{1.1}$$

subject to the constraints

$$y_i\left(w^r \emptyset(x_i) + b\right) \geq 1 - \xi_i \text{ and } \xi_i \geq 0, i = 1, \ldots, N \tag{1.2}$$

Note that $y \in \pm 1$ is the class label and x_i is the independent variables.

$$\frac{1}{2}w^r w - vp + \frac{1}{N}\sum_{i=1}^{N} \xi_i \tag{1.3}$$

subject to the constraints

$$y_i\left(w^r \phi(x_i) + b\right) \geq \rho - \xi_i, \xi_i \geq 0, i = 1, \ldots N \text{ and } \rho \geq 0 \tag{1.4}$$

CNN Classifier

CNN is a text classifier using feed-forward artificial neural networks and uses a different multilayer perceptron model to involve minimal preprocessing. The outcome of each convolution wills flames when a particular pattern is identified. As a result of varying the size of the kernels and concatenating their outcome, its permits to detect patterns of various sizes containing two, three, or five adjacent words. The output patterns could be in terms like ‘I hate’, ‘very good’ and consequently CNNs can categorize them in the sentence in spite of their position.

Performance Analysis

The performance analysis of CNN with Naive Bayesian classifiers, SVM classifiers is carried out. Then the evaluated performance metrics are listed as follows (Table 1.1).

a. Sensitivity

Table 1.1 Efficiency of character recognition methods

Metrics	Naive Bayesian classifiers	SVM classifiers	CNN classifiers
Sensitivity	83.66	89.13	91.13
Specificity	84.42	90.63	92.63
Accuracy	86.56	91.05	93.45

b. Specificity
c. Accuracy.

Mean Square Error

MSE calculates the difference between the pixel values of the original image and the recognized image. So, the MSE is calculated by using following Equation.

$$\mathrm{MSE} = \frac{1}{\mathrm{MN}} \sum_{i=1}^{M} \sum_{j=1}^{N} \left(f(i, j) - f'(i, j)\right)^2 \tag{1.5}$$

In Eq. (1.5) $f(i, j)$ is represented as the original image and $f^{'}(i, j)$ is denoted as the recognized character image. M is the height of the image and N is the width of the image.

Unicode Mapping

The Unicode standard reflects the basic principle which emphasizes that each character code has a width of 16 bits. Unicode text is simple to parse and process Unicode characters have well-defined semantics. Hence, Unicode is chosen as the encoding scheme for the current work. After classification, the characters are recognized and a mapping table is created in which the Unicode for the corresponding characters are mapped. Based on the mapping, the Sanskrit characters are recognized with minimum error rate also enhances the recognition accuracy.

Results and Discussion

The efficiency of the CNN network is analyzed using the sensitivity and specificity metrics Table 1.2 clearly explains that the CNN method consumes minimum error rate while classifying the Sanskrit characters when compared to the existing method. The minimum error rate leads to increase the accuracy of the CNN system. From the above discussions, the proposed system classifies the Sanskrit characters with efficient manner when compared to the other traditional methods.

Table 1.2 Mean square error of CNN

Different classifiers	Mean square error
Naïve Bayesian classifier	0.9
SVM classifier	0.789
CNN classifier	0.645

Conclusion

The system explains the various processes of Sanskrit character recognition process namely preprocessing, segmentation, feature extraction and classification. The character recognition system methods are analyzed and the several quality measures such as sensitivity, specificity and accuracy have been used to analyze the effectiveness of the proposed techniques. Thus the CNN-based recognition system recognizes the exact handwritten characters in offline with minimum error rate and high accuracy when compared to the traditional classifiers. Thus, the CNN classifier overcomes the above-discussed classifiers' drawback with minimum error and high efficiency.

References

1. Agarwal, P.: Hand-written character recognition using Kohonen network. IJCST **2**(3), 10–18 (2011)
2. Alirezaee, S., Aghaeinia, H., Faez, K., Fard, A.S.: An efficient feature extraction method for the middle-age character recognition. Lect. Notes Comput. Sci. **3645**(PART II), 998–1006 (2005)
3. Ameri, M.R., Haji, M., Fischer, A., Ponson, D., Bui, T.D.: A feature extraction method for cursive character recognition using higher-order singular value decomposition. In: 2014 14th International Conference on Frontiers in Handwriting Recognition, pp. 512–516 (2014)
4. Dwivedi, N., Srivastava, K., Arya, N.: Sanskrit Word Recognition Using Prewitt's Operator and Support Vector Classification. ICECCN, pp. 265–269 (2013)
5. El Qacimy, B., Kerroum, M.A., Hammouch, A.: Feature extraction based on DCT for handwritten digit recognition. Int. J. Comput. Sci. Issues (IJCSI) **11**(6), 27–27 (2014)
6. Joshi, M.R., Sabale, M.V.V.: Offline Character Recognition for Printed Text in Devanagari Using Neural Network and Genetic Algorithm, pp. 1–8 (2013)
7. Kale, K.V., Deshmukh, P.D., Chavan, S.V., Kazi, M.M., Rode, Y.S.: Zernike moment feature extraction for handwritten Devanagari compound character recognition. Sci. Inf. Conf. **3**(1), 459–466 (2013)
8. Liu, X.Y., Blumenstein, M.: Experimental analysis of the modified direction feature for cursive character recognition. In: Proceedings—International Workshop on Frontiers in Handwriting Recognition, IWFHR, pp. 353–358 (2004)
9. Dineshkumar, R., Suganthi, J.: Sanskrit character recognition system using neural network. Indian J. Sci. Technol. **8**(1), 65–69. E-ISSN: 0974-5645 (2015)
10. Dineshkumar, R.: Offline Sanskrit handwritten character recognition framework based on multilayer feed forward network with intelligent character recognition. Asian J. Inf. Technol. **15**(11), 1678–1685. ISSN: 1682-3915 (2016)

Chapter 2
Resampling Imbalanced Data and Impact of Attribute Selection Methods in High Dimensional Data

K. Ulaga Priya and S. Pushpa

Introduction

In recent years, many real-world problems are characterized by imbalanced data where the distribution of classes is skewed. A dataset is imbalanced [1] when the classification categories are not approximately equivalent, for example one class may be extremely low (minority class) and the other class may be extremely high (majority class). This imbalanced nature of data leads the prediction algorithm to be biased toward the minority class. The poor representation of minority class affects the performance of the classification algorithm, which is evident through various assessment metrics. Applying a resampling algorithm and balancing the data resolves this issue. Studies reveal that apart from resampling technique using feature selection method improves the performance tremendously.

Current Approach

The current approach carried out in this paper for highly imbalanced data is applying feature selection technique and subsequently apply the resampling technique. The sampling technique chosen for this experiment is Random under sampling and several feature selection techniques were applied on each dataset.

K. Ulaga Priya (✉)
St. Peter's Institute of Higher Education and Research, Chennai, India
e-mail: ulagapriya@gmail.com

K. Ulaga Priya · S. Pushpa
Vels Institute of Science Technology and Advanced Studies, Chennai, India

Ch. Satyanarayana et al. (eds.), *Proceedings of the International Conference on Computer Vision, High Performance Computing, Smart Devices and Networks*, Advanced Technologies and Societal Change,
https://doi.org/10.1007/978-981-19-4044-6_2

Re-sampling Method

Several resampling technique is available in literature namely Random under Sampling (RUS), Random Over Sampling (ROS), Synthetic Minority Oversampling Technique (SMOTE), ADaptive SYNthetic sampling (ADASYN), Random Under Sampling (RUS), CONDENSED NEAREST NEIGHBOR (CNN), Edited Nearest Neighbor (ENN), and Neighborhood Cleaning Rule (NCL). This paper uses Random under sampling technique which under samples the majority data [2]. The algorithm randomly selects the majority instances which are equivalent to the number of instances in minority class. With this the class distribution can be more balanced, however, important information may be lost when examples are eliminated at random. In [3] the data is balanced and the classification would yield better results.

Feature Selection

Feature selection [4] is an important technique used in high dimensional data where irrelevant and redundant data is removed and the dimensionality is reduced. Classification will be performed with the selected attribute provided by any feature selection technique. Several feature selection techniques are available namely Information gain, Random forest selector, Lasso regression, Forward feature selection, Backward feature selection, and combination of forward and backward feature selection.

- Filter Method
  - Information gain
- Wrapper Methods
  - Forward stepwise Selection
  - Backward stepwise Selection
- Stepwise Selection
- Embedded Methods
  - LASSO
  - Random Forest

Information Gain

This is a feature selection technique using Filter method. Information gain tells us how much information [5] is given by the independent variable about the dependent variable Information gain is helpful in case of both categorical and numerical dependent variables. For numeric dependent variables, bins are created. Information gain is the amount of information that is gained by knowing the value of the attribute which is the entropy of the distribution. Information gain = (Entropy of distribution before the split)-(entropy of distribution after it)

$$\mathrm{IG}(T, c) = H(T) - H(T|c)$$

Forward Stepwise Selection

The algorithm begins with an empty model and it keeps adding the important variables one by one to the [6] model. In other words it starts with a model containing no predictors, and then adds predictors to the model, one-at-a-time, until all of the predictors are in the model. The important variable is added at each step to the model based on smallest *p*-value or increase in R2.

Backward Stepwise Selection

This algorithm starts with full model which contains all variables under consideration. Then the least important variables [7] are removed from the model until the threshold value is reached. The threshold value is determined by AIC and BIC.

Stepwise Selection

This is a hybrid approach where forward stepwise [3] selection and backward stepwise selection are combined together. In this algorithm variables are added sequentially to the model in analogy with forward selection. However, after adding each variable this algorithm may also remove variable which no longer provide improvement in analogy with backward selection.

Lasso

LASSO is a powerful technique which performs two [8] main tasks; regularization and feature selection. The method shrinks (regularizes) the coefficients of the regression model as part of penalization. For feature selection, the variables which are left after the shrinkage process are used in the model. Like every other algorithm goal here is to minimize the prediction error.

Random Forest

In every tree grown in the forest, put down the oob [9] cases and count the number of votes cast for the correct class. Now randomly permute the values of variable m in the oob cases and put these cases down the tree. Subtract the number of votes for the correct class in the variable-m-permuted oob data from the number of votes for the correct class in the untouched oob data. The average of this number over all trees in the forest is the raw importance score for variable m.

Experimental Setup

Datasets

The data is taken from KEEL and openML websites. The data is taken such a way that they are imbalanced and has lot of features.

- wall-robot-navigation1_1
- thyroid
- vehicle
- cardiotocography
- wine.

Design of Experiments

- Apply the feature selection algorithm and select the important variables.
- The data set should splitted into k folds, each one containing 20% of the patterns of the dataset.
- For each fold, data is splitted into training set and test set.
- The training data is either under sampled to generate a balanced training sample.
- The decision tree classifier should be applied on the balanced training data with selected features.
- The performance of the classifier is evaluated against the test set.
- The above steps should be repeated for different sampling technique and evaluate the best results.

Result and Discussion

A fivefold cross validation for four times was executed to get the performance metrics for 20 runs. On each fold data is splitted into training set and test set. The training data set is under sampled by Random under sampling (RUS) and [10] the balanced data is used by the decision tree classifier with selected features given by feature selection algorithm. The classifier performance is evaluated against the test set in each fold. The performance measures precision, Recall, AUCROC is calculated for each feature selection method. The results are tabulated for each dataset in Table 2.1 for sample datasets cardiotocography and vehicle. Similarly, other datasets were also experimented.

The experiment study evaluates four performance metrics than accuracy which are more suitable for Imbalanced data. Many researchers adhere to one metric but several metrics are available there is no best metric when considering Imbalanced data. False positive rate, False Negative rate and True positive rate and True Negative rate are used in the evaluation metric precision, recall, and F1score. AUCROC curve is another metric where it draws ROC curve with true positive rate on y axis and false positive rate on x axis.

Recall = True Positive /(True Positive + False Negative)
Precision = True Positive/(True Positive + False Positive)
AUCROC indicates how well the probabilities from the positive classes are separated from the negative classes.

Results reveal that after applying feature selection techniques in imbalanced data, there is promising result with high performance. Information gain and random forest gives the highest performance when compared to other Feature selection techniques in Table 2.1.

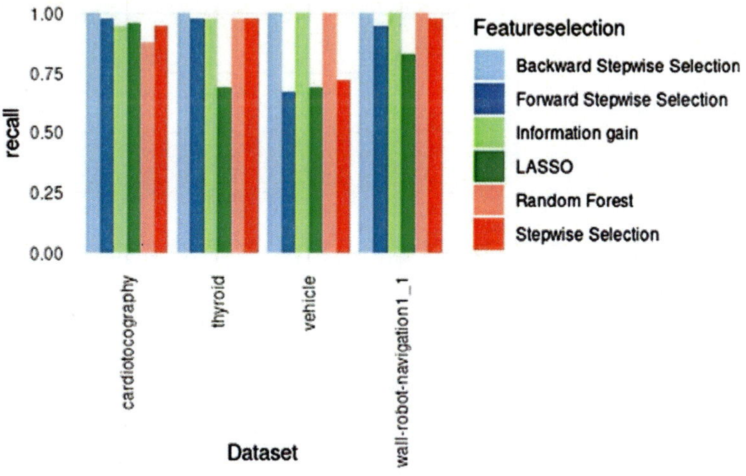

Table 2.1 Results

Feature selection	Dataset	No_of_features	Precision	Recall	AUCROC
Information gain	wall-robot-navigation1_1	25	1	1	1
Random forest	wall-robot-navigation1_1	25	1	1	1
LASSO	wall-robot-navigation1_1	25	0.81	0.83	0.89
Forward stepwise selection	wall-robot-navigation1_1	25	0.94	0.95	0.93
Backward stepwise Selection	wall-robot-navigation1_1	25	1	1	1
Stepwise selection	wall-robot-navigation1_1	25	0.97	0.98	0.98
Information gain	Thyroid	21	0.79	0.98	0.99
Random forest	Thyroid	21	0.77	0.98	0.99
LASSO	Thyroid	21	0.44	0.69	0.82
Forward stepwise Selection	Thyroid	21	0.84	0.98	0.99
Backward stepwise Selection	Thyroid	21	1	1	1
Stepwise Selection	Thyroid	21	0.74	0.98	0.98
Information gain	Vehicle	19	1	1	1
Random Forest	Vehicle	19	1	1	1
LASSO	Vehicle	19	0.67	0.69	0.86
Forward Stepwise Selection	Vehicle	19	0.66	0.67	0.82
Backward Stepwise Selection	Vehicle	19	1	1	1
Stepwise Selection	Vehicle	19	0.68	0.72	0.85
Information gain	Cardiotocography	36	0.89	0.95	0.97

(continued)

Table 2.1 (continued)

Feature selection	Dataset	No_of_features	Precision	Recall	AUCROC
Random Forest	Cardiotocography	36	0.97	0.88	0.94
LASSO	Cardiotocography	36	0.95	0.96	0.98
Forward Stepwise Selection	Cardiotocography	36	0.9	0.98	0.97
Backward Stepwise Selection	Cardiotocography	36	1	1	1
Stepwise Selection	Cardiotocography	36	0.91	0.95	0.97

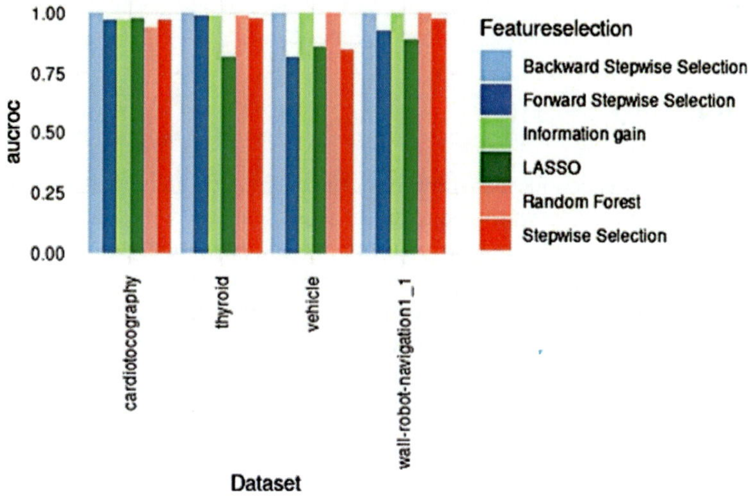

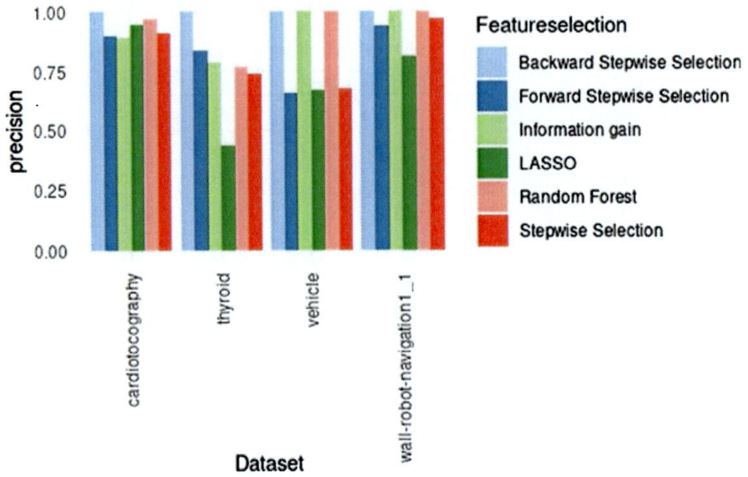

Conclusion

The experiment was conducted with 6 imbalanced datasets and 5 feature selection techniques. The results are promising and it is evident that with feature selection technique on imbalanced data data performs well. Information gain and Random forest provides highest performance across the datasets. The key finding from this paper is that feature selection is beneficial to handling most high-dimensional imbalanced data sets. Future research should explicitly compare and contrast the size, skew, noise, and dimensionality of imbalanced data sets to determine the influence each of these characteristics has on the performance of resampling methods, algorithms, and feature selection methods.

References

1. Deng X, Zhong W, Ren J, Zeng D (2016) ANI Mbalanced data classification method based on automatic clustering under - Sampling
2. López, V., Fernández, A., García, S., Palade, V., Herrera, F.: An insight into classification with imbalanced data: empirical results and current trends on using data intrinsic characteristics. Inf. Sci. (Ny) **250**, 113–141 (2013). https://doi.org/10.1016/j.ins.2013.07.007
3. Singh, R., Raut, R.: Review on class imbalance learning: binary and multiclass. Int. J. Comput. Appl. **131**(16), 4–8 (2015). https://doi.org/10.5120/ijca2015907573
4. Lopez, K.M.M., Magboo, M.S.A.: A clinical decision support tool to detect invasive ductal carcinoma in histopathological images using support vector machines, Naïve-Bayes, and knearest neighbor classifiers. Front. Artif. Intell. Appl. **332**(Idc):46–53. https://doi.org/10.3233/FAIA200765

5. Al Khaldy, M.: Resampling imbalanced class and the effectiveness of feature selection methods for heart failure dataset. Int. Robot. Autom. J. **4**(1), 37–45 (2018). https://doi.org/10.15406/iratj.2018.04.00090
6. Wang, S., Yao, X.: Multiclass imbalance problems: analysis and potential solutions. IEEE Trans. Syst. Man, Cybern. Part B Cybern. **42**(4), 1119–1130 (2012). https://doi.org/10.1109/TSMCB.2012.2187280
7. Deng, H., Runger, G.: Gene selection with guided regularized random forest. Pattern Recognit. **46**(12), 3483–3489 (2013). https://doi.org/10.1016/j.patcog.2013.05.018
8. Lango, M.: Tackling the problem of class imbalance in multi-class sentiment classification: an experimental study. Found. Comput. Decis. Sci. **44**(2), 151–178 (2019). https://doi.org/10.2478/fcds-2019-0009
9. Wainer, J., Franceschinell, R.A.: An empirical evaluation of imbalanced data strategies from a practitioner's point of view arXiv **Ic** (2018)
10. Li, H., Zou, P., Han, W., Xia, R.: A combination method for multi-class imbalanced data classification. In: Proceeding—2013 10th Web Information Systems and Applications Conference WISA 2013, no. 1, pp. 365–368 (2013). https://doi.org/10.1109/WISA.2013.75

Chapter 3
Wireless sEMG Acquisition and Monitoring—A Survey

R. D. Delina Rajkumari, Prakash Rajiah, A. Lakshmi Sangeetha, A. Pravin Renold, and A. Balaji Ganesh

Introduction

The recent advancements in Electronics in medical field have paved the way for e-health and mobile health systems, electronic health record systems, remote patient monitoring, etc. These systems are used for continuous monitoring, prediction, diagnosis and treatment. Hence they help in reducing the cost of utilizing the facility, reduce the time taken and enable us to monitor the vital signs while performing the daily routine [1].

EMG sensors are widely used in medical field to obtain information about the activity of muscles. These signals are obtained from a tool called electromyograph. These signals are much helpful in diagnosing neuromuscular diseases such as Parkinson's disease, etc. But the application is not only limited in diagnosis of diseases. The applications also include medical research, ergonomics and rehabilitation, biomechanics and sports science applications [2, 3]. The analysis of muscle activity can give knowledge about the muscle fatigue conditions. The EMG biofeedback is capable of providing real-time information about the skeletal muscle activity so that the muscle performance could be easily understood. This can be used in muscle fatigue detection.

R. D. Delina Rajkumari (✉) · A. Lakshmi Sangeetha · A. Pravin Renold · A. Balaji Ganesh
Electronic System Design Lab, Velammal Engineering College, Chennai 600066, India
e-mail: delinarajkumari@velammal.edu.in

A. Balaji Ganesh
e-mail: abganesh@velammal.edu.in

P. Rajiah
Emsensing Technologies Limited, Chennai, India

Ch. Satyanarayana et al. (eds.), *Proceedings of the International Conference on Computer Vision, High Performance Computing, Smart Devices and Networks*, Advanced Technologies and Societal Change,
https://doi.org/10.1007/978-981-19-4044-6_3

Fatigue is a natural outcome of long-term activities [4, 5]. Studies on gym and yoga overtraining in individuals indicate that fatigue occurs in muscles prior to injuries that can lead to series of complications [6, 7].

Patients, who have experienced fractures or muscle injury, undergo physiotherapy, yoga therapy or isokinetic exercises for functional rehabilitation [8]. The muscles undergo dynamic contraction during these processes. However, if the exercises are done excessively or incorrectly, it may worsen the condition. Hence the EMG monitoring during exercise/yoga therapy is important for sick as well as healthy people [2].

Commercial EMG tools are costly and require high maintenance. Also, the wired EMG acquisition systems cause discomfort while the patients carry on with their routine activities or during therapies. Hence there is a need for development of wireless EMG acquisition systems [9].

This paper presents a comprehensive survey of the recent works addressing remote EMG monitoring systems.

Literature Survey

The EMG signals can be acquired by invasive and non-invasive electrodes. The invasive method involves inserting needle electrodes into the muscle fibre, which causes discomfort while performing the exercise. Hence Surface EMG (sEMG) methods are preferred, where electrodes are placed on the surface of the skin and the electrical activity of the muscle is monitored.

The wireless EMG acquisition system usually consists of the EMG electrodes, Signal conditioning unit, the Microcontroller Unit (MCU) and the transceiver unit to transmit the measured EMG signal.

According to the survey, different transceiver modules were used for wireless EMG acquisition systems which predominantly consisted of Bluetooth, Wi-Fi, GSM/GPRS, X-bee, etc.

Bluetooth-Based Wireless EMG Sensor

Heaffey et al. [10] developed a low-power wearable device for sEMG and muscle fatigue monitoring. The wearable platform integrates user interface and application-specific Integrated circuits that can estimate the Muscle Fibre Conduction Velocity (MFCV) which in turn is used to analyze the state of muscle fatigue. The system also consisted of a BLE System on chip that can calibrate the muscle fatigue and relay the information to android application. The Application-Specific Integrated Circuit (ASIC) consists of three conductive and one reference electrode. nRF51822 is used for ADC conversion and to transmit the data via Bluetooth. Through the android application, the user can set the comparator value and other parameters and

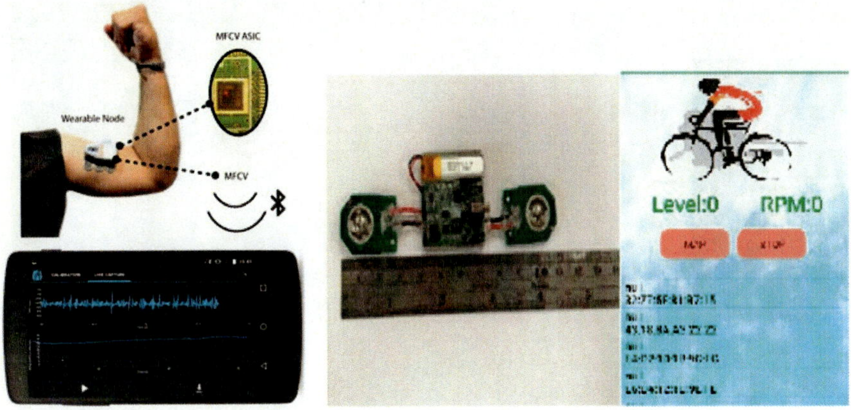

Fig. 3.1 sEMG sensor and front end of Bluetooth-based sensor [10, 11]

the EMG data can be visualized. The system was tested on Biceps Brachii of two test subjects.

Liu et al. [11] proposed a method of developing an EMG patch that could monitor the muscle performance while exercising. ARM-Cortex M4 is the microcontroller unit used and the real-time data is transmitted using BLE module. The MCU is used to measure the median frequency of Empirical Mode Decomposition (EMD) for the Surface EMG signal. In the mentioned work, the computer-based measurement system was designed to analyze the MF values of sEMG signal offline. An APP was designed to monitor the muscle fatigue and the riding information of the user. Figure 3.1 shows the surface EMG sensor and the front end of the developed mobile application.

Wi-Fi-Based Wireless EMG Sensor

Yang et al. [12] proposed a low-cost wireless multichannel sEMG acquisition system that focuses on high sampling rate and multichannel data acquisition. The system is divided into two modules—the measurement module and host. The measurement module measures the analogue EMG data and transmits it to the host. The host receives and decodes the data to achieve multichannel detection. The weak EMG signals are amplified, filtered and converted to digital signals and are sent to Wi-Fi module through UART interface. The performance of the system is compared with a commercial sensor and the co-relation coefficients represent moderate to high relative agreement between the proposed and commercial system.

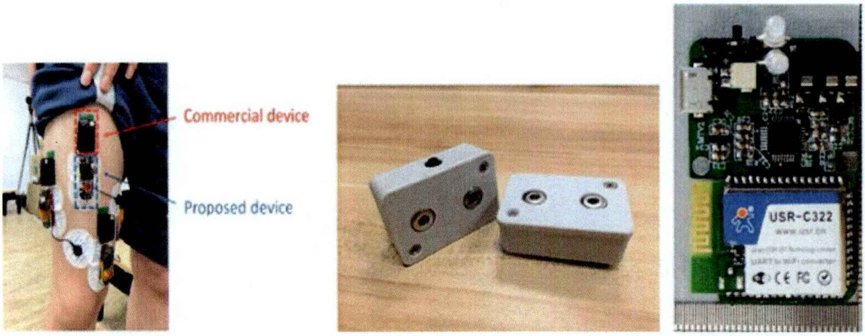

Fig. 3.2 sEMG sensor of Wi-fi based devices [12, 13, 14]

Wu et al. [13] proposed a low-cost EMG sensor for hand motion recognition where the raw EMG signals are used for training the model and then to the online gesture recognition. The sensor circuit consists of electrode snap fasteners, amplifiers, filters and level rising circuits. C8051F320 is the MCU and USR-C322 is the Wi-Fi module. The amplified signal is digitalized using C8051F320 and sent over the Wi-Fi. The model training is done using MATLAB and visualization is done using the software.

Wang et al. [14] proposed the design of Wireless Personal Area Network (WPAN) of ECG/EMG sensors that offers multi-node connectivity using CyberSens—developed in the Human Cyber-Physical Systems Laboratory, Florida International University. The data are sent over Wi-Fi using built-in Wi-Fi module of CC3200. The system is also provided with Inertial Measurement Unit (IMU) and onboard data storage. A graphic user-interface software package was designed in the MALAB environment for monitoring the signals and the circuit is powered by 400 mAH battery. Figure 3.2 shows the Wifi based surface EMG sensors that were developed.

X-Bee-Based Wireless EMG Sensor

Ishak et al. [15] developed an X-bee-based wireless EMG monitoring. The muscle actions were captured using commercial EMG electrode. The signal is then amplified and filtered for unwanted noise and transmitted using X-bee module to the computer as GUI captured wireless signals. The spectral analysis is done using FFT filter. The NI6009 DAQ card was utilized for the A/D converter while graphic user-interface software (LABVIEW) was employed to view and record the captured signals.

Fig. 3.3 sEMG module of GSM based sensor [16]

Other Wireless EMG Sensor

Sayyed et al. [16] proposed an HTTP protocol-based EMG monitoring system. The signals are captured using suction electrodes, amplified, filtered and converted into digital signals and transmitted to remote server using HTTP protocol. The weak EMG signals are amplified using AD620. LPC1768 is the MCU used and it acquires the samples and stores them in memory. The samples are transmitted using SIM900A. The transmitted signals are captured using GPRS, stored in the server and displayed on web browser.

Kledrowetz [17] provided a low-power miniaturized solution where an ASIC-based sensor device was developed. The electrodes are connected to the ASIC which is composed of Instrumentation amplifiers, filters and Delta- Sigma modulator which is used to convert the amplified signal to pulse density modulated (PDM) signal using a 12-bit delta-sigma modulator. The ASIC is designed and fabricated in TSMC0.18 mixed-signal CMOS technology. RF module is used to send data to logger (PC). The system is powered by 3v button cell. The developed module is shown in Fig. 3.3.

Comparison of the Existing Systems

In [10], Battery life with BLE inactive is 64 h. 10-bit ADC LSB is 1.172 mV. EEPROM fatigue data capacity 130816 values. Three EMG electrodes and one reference electrode were used.

In [11], Wireless method was used to charge the lithium battery. The EMD algorithm was used to obtain enveloped sEMG signal. Two electrodes were used to obtain the EMG signal. The fatigue was monitored using an APP built on JavaScript.

In [12], the raw EMG signals were passed through band filter (50–200 Hz) and the power line noises were eliminated using notch filter (55–65 Hz). It was compared with commercial device TrignoTM Wireless Systems and Smart Sensors, Delsys, Inc., USA. The SNR is found to be 34 which is less than the commercial device. 300 mAh lithium battery is used as power source.

In [13], four muscle sensors were used for gesture recognition. A 3.7v, 300 mAH battery is used as the power source. It provided a low-cost solution.

The [14] proposed a WPAN where multiple sensors can be used to obtain the signal. To support high data rate, UDP protocol was used. It also provides onboard battery charging.

In [15], the RC band pass filtered signal is used for FFT analysis. The average SNR was found to be 57.5. In the work, the SNR increased by applying Butterworth band pass filter. The signal quality and accuracy increased with the application of digital filter.

In [16], three electrodes are used to obtain the sEMG signals. AD620 is used as Instrumentation amplifier. The electrical activity of the muscle was captured, conditioned and transmitted using SIM900A and also visualized using digital oscilloscope.

In [17], the overall power consumption is found to be 4.4 mW. The area is 2.56 mm^2. It provides a low-cost, miniaturized solution and 6–7 battery life while sending data.

Summary

The comparison of the developed modules from the literature are shown in Table 3.1.

Conclusion

In the survey paper, four different implementations of wireless sEMG acquisition systems were discussed along with their design, merits and demerits. Low-power solutions are provided by [10, 14, 16]. The high data rates are provided by [12, 14]. Low-cost solutions are provided by [12]. Compact designs are proposed in [10, 13, 14]. ASIC is developed in [10, 17]. The ideas may be used to develop a more efficient sEMG acquisition and processing system for monitoring and fatigue detection.

Table 3.1 Comparison of the existing systems

	Pre-amplifier	Pre-processing	Processing unit	Visualization	Data storage	App development	Web/Software development	Power source	Electrodes used	Filter used
Heaffey et al. [10]		✓	nRF51822	✓	✓	✓		Rechargeable Lithium polymer	3 + 1	
Liu et al. [11]		✓	ARM-CORTEX M4	✓		✓		Rechargeable Lithium	2 + 1	Butterworth
Yang et al. [12]	✓	✓	ARM-CORTEX M0					Lithium battery	2 + 1	IIR notch
Ishak et al. [15]		✓	NI6009 DAQ	✓	✓					Butterworth
Sayyed et al. [16]		✓	ARM-CORTEX M3 LPC1768	✓			✓	9v battery	Suction cup electrode	
Kledrowetz et al. [17]		✓	ATmega88					LR44	2 + 1	
Wu et al. [13]		✓	C8051F320	✓			✓	Rechargeable Lithium	2	Notch
Wang et al. [14]		✓	ADS1292	✓	✓		✓	Polymer Lithium ion		

Acknowledgements The study was a part of the project funded by Department of Science and Technology. The authors acknowledge the financial support from Science and Technology of Yoga and Meditation (SATYAM) under the Department of Science and Technology, New Delhi, India for sanctioning the project-File No.: DST/SATYAM/2018/20(G) to Velammal Engineering College, Chennai.

References

1. Rahim, A., Forkan, M., Khalil, I.: A probabilistic model for early prediction of abnormal clinical events using vital sign correlations in home-based monitoring. In: Proceedings of the 2016 IEEE International Conference on Pervasive Computing and Communications, pp. 14–19, Sydney, NSW, Australia (2016)
2. Halabi, R., El Banna, I., Malaeb, R., Halabi, R., Diab, M.: Novel approach for wireless EMG database collection: applied to muscle building workout routine optimization. In: Fifth International Conference on Advances in Biomedical Engineering (ICABME), pp. 1–4 (2019)
3. Stastny, P., Gołaś, A., Blazek, D., Maszczyk, A., Wilk, M., Pietraszewski, P., Petr, M., Uhlír, P., Zając, A.: A systematic review of surface electromyography analyses of the bench press movement task. PLoS ONE **12** (2017)
4. Opar, D.A., Williams, M.D., Shield, A.J.: Hamstring strain injuries: Factors that lead to injury and re-injury. Sports Med. **42**(3), 209–226 (2012)
5. Enoka, R.M., Duchateau, J.: Translating fatigue to human performance. Med. Sci. Sports Exerc. **48**(11), 2228–2238 (2016)
6. Mueller-Wohlfahrt, H.W., Haensel, L., Mithoefer, K., Ekstrand, J., English, B., McNally, S., Orchard, J., van Dijk, C.N., Kerkhoffs, G.M., Schamasch, P., Blottner, D., Swaerd, L., Goedhart, E., Ueblacker, P.: Terminology and classification of muscle injuries in sport: the Munich consensus statement. Br. J. Sports Med. **47**(6), 342–350 (2013)
7. Kellmann, M.: Preventing overtraining in athletes in high-intensity sports and stress/recovery monitoring. Scand. J. Med. Sci. Sports **20**(2), 95–102 (2010)
8. Gränicher, P., Stöggl, T., Fucentese, S.F., et al.: Preoperative exercise in patients undergoing total knee arthroplasty: a pilot randomized controlled trial. Arch. Physiotherapy **10**, 13 (2020)
9. Mabrouk, M.S., Kandil, O.A.: Surface multi-purposes low power wireless electromyography (EMG) system design. Int. J. Comput. Appl. **41**(12), 10–16 (2012)
10. Heaffey, J., Koutsos, E., Georgiou, P.: Live demonstration: wearable device for remote EMG and muscle fatigue monitoring. In: 2015 IEEE Biomedical Circuits and Systems Conference (BioCAS), pp. 1–5 (2015)
11. Liu, S., Lin, C.-B., Chen, Y., Chen, W., Huang, T.-S., Hsu, C.-Y.: An EMG patch for the real-time monitoring of muscle-fatigue conditions during exercise. Sensors **19**(14), 3108 (2019)
12. Yang, Y.-H., Ruan, S.-J., Chen, P.-C., Liu, Y.-T., Hsueh, Y.-H.: A low-cost wireless multichannel surface EMG acquisition system. IEEE Consum. Electron. Mag. **9**(5), 14–19 (2020)
13. Wu, C., Yan, Y., Cao, Q., Fei, F., Yang, D., Song, A.: A low cost surface EMG sensor network for hand motion recognition. In: 2018 IEEE 1st International Conference on Micro/Nano Sensors for AI, Healthcare, and Robotics (NSENS), pp. 35–39 (2018)
14. Wang, Q., et al.: A high data rate, multi-nodes wireless personal-area sensor network for real-time data acquisition and control. In: First International Conference on Electronics Instrumentation & Information Systems (EIIS), pp. 1–5 (2017)

15. Ishak, A.J., Ahmad, S.A., Soh, A.C., Naraina, N.A., Jusoh, R.M.R., Chikamune, W.: Design of a wireless surface EMG acquisition system. In: 24th International Conference on Mechatronics and Machine Vision in Practice (M2VIP), pp. 1–6 (2017)
16. Sayyed, R., Akhter, N., Khan, A., Rabbani, G.: Remote monitoring of EMG signals. Int. J. Res. Anal. Rev. **6**(2), 135–141 (2019)
17. Kledrowetz, V., Prokop, R., Fujcik, L., Pavlík, M., Háze, J.: Low-power ASIC suitable for miniaturized wireless EMG systems. J. Electr. Eng. **70**, 393–399 (2019)

Chapter 4
Black Hole Algorithm for BigData Anonymization

U. Selvi and S. Pushpa

Introduction

Over the decades, Hardware and software have had a rapid growth toward the advancement of storage capability. All personal information regarding an individual is available online and usage of Databases has been booming. Small set of data can be easily handled by the Data Mining tools available but handling large set of data is a challenging task. Preserving privacy is another challenging issue when dealing with sensitive data. When dealing with data using traditional tools like relational Databases, data is found in the term of tuples where we have attributes that describe individuals. Four types of attributes that explicitly describes the individuals are as follows: explicit identifiers1, quasi-identifiers2, sensitive attributes, and non-sensitive attributes [1]. The widespread methodology for conserving privacy is anonymization which is to hide the explicitly identifying individual attributes and standard Algorithm for this is k-anonymity. But though such approach seems to be simple in implementation, which is not sufficient and individuals can be re-identified.

The flaw of straightforward k-anonymity was revealed [2, 3] and was further confirmed by de Montjoye et al. in [4]. By combining two datasets, L. Sweeney was successful in identifying the individuals by attack named "linking attack". K-anonymity was proposed to address those attacks and it becomes the base algorithm for the next subsequent algorithm related to anonymization. In k-anonymity, each record cannot be renowned among a minimum of k-1 records. Throughout this process, dataset is divided into several groups based on similarity classes and the proceedings of each group are generalized. Hence it becomes difficult to spot individuals in group, since all the individuals of equivalent groups are similar. Thus

U. Selvi (✉) · S. Pushpa
Department of Computer Science and Engineering, St. Peter's Institute of Higher Education and Research, Avadi, Chennai 600054, India
e-mail: slvunnikrishnan@gmail.com

Ch. Satyanarayana et al. (eds.), *Proceedings of the International Conference on Computer Vision, High Performance Computing, Smart Devices and Networks*, Advanced Technologies and Societal Change,
https://doi.org/10.1007/978-981-19-4044-6_4

the objective of anonymization is achieved. This approach seems to be similar to clustering-based approach in which each equivalence class is grouped as cluster. Clustering-based k-anonymity has a data utility as they group related records organized. Although the k-anonymity constructed clustering is theoretically simple, the computational complication of discovering an optimal k-anonymous solution is NP-hard [5]. In this context, great effort is required to provide a complete search for optimal solutions. In this case, it should be feature with minimum information loss. However, this process fails when the dataset increases and data suffer bad data quality. Many meta-heuristics approaches were found to be effective in those areas but more exploration is needed in terms of privacy and anonymity.

Meta-heuristic algorithms are optimization methods but require expensive computation time. Meta-heuristic algorithms are simple to implement and have a simple structure and reduced number of parameters; it is called as region Algorithm (BHA) and is free from parameter setting issues. BHA algorithm has never been applied to the problem of privacy-preserving and anonymization. BHA algorithm is summarized as follows: It is a population-centered algorithm that has the region of space. It has a gravitational force in which any object in the universe gets disappeared if it gets close to it. The BHA Algorithm [6] applied to k-anonymity problem signifies the k-anonymous solution and the top explanation given by black hole. The algorithm starts with an initial population of candidate clarifications produced accidentally. Its objective is to select the optimal k-anonymity result which has the minimum information loss. To enhance info quality, clustering algorithm group the similar quasi-identifiers within the group having a minimum of k records. The similarity is calculated supported information loss as a distance and cost metric. This makes sure that fewer misrepresentations are required to anonymize the record in a cluster which enhances data quality.

The rest of the paper is systematized as follows. Subsequent segment surveys correlated work around k-anonymity-centered approaches. Our algorithm is presented in Sect. 4.3 and is experimentally assessed in Sect. 4.4. We determine this paper in Sect. 4.5.

Related Work

k-anonymization becomes the standard Benchmark and base algorithm for much privacy-preserving algorithms. Since our proposed work is around k-anonymity, some literature survey is done around k-anonymity and cluster algorithm.

Anonymity based on clustering approaches

Clustering-based anonymization on attributes hierarchies by local recording was proposed [7]. Equivalence class was created and this approach tries to select the correspondence class of size lesser than k. It then calculates the space among C and catches the similarity C′ with the small distance to C. Lastly the two similar class is group and generalized. The process is repeated until the equivalence class has a minimum distance of k records. Weighted Feature C-means clustering for k-anonymity was projected [8]. The process begins as follows: C random records were selected as seeds and by calculating the number of equivalence classes. Then the algorithm starts to assign weights to each quasi-identifier. The process continues to identify records close to the seeds and feature weight is updated to reduce information loss. This process is iterated until no change is applied to the clusters of record. The algorithm merges small equivalence class which has k records with larger correspondence classes to satisfy the k-anonymity constraint.

K-member clustering was projected by Byun et al. [9]. The algorithm tries to build a cluster selecting record around the seed and form k-1 nearest record. Then, the algorithm selects the replacement record for the record farther from the seed and iterates the process to create a cluster. And also assigns the unassigned record to any closer clusters. The process ends when all the records are assigned. Greedy k-anonymity algorithm was proposed by Loukides and Shao [10]. This is similar to k-member clustering but differs from assigning cluster to a group awaiting a user demarcated threshold is extended. The cluster which has record lesser than 'k' records will be deleted.

One pass k-means Algorithm (OKA) was projected by Lin and Wei [11] to achieve anonymity-based clustering. During the first phase, k-mean algorithm was applied and during the second phase, clusters having records more than k records are adjusted by moving records to cluster having less than k records. Clustering-primarily based K-anonymity with a set of rules known as GCCG was suggested by Ni et al. [12]. This methodology is composed of four steps namely Grading, Centering, Clustering, and Generalization. In Grading and Centering steps, the facts are looked after primarily constructed totally at the rating computed of every file then the primary X facts are selected as centroids. The next stage is foundation of clusters through including to every centroid the k-1 closest facts. In the very last phase, the facts are generalized. To decorate the overall piece of GCCG set of rules, the authors additionally suggest a parallelized model of GCCG.

A clustering-primarily based totally k-anonymity set of rules which deliberates the general delivery of quasi-identifier businesses in a multidimensional space was projected by Zheng et al. [13]. The proposed set of rules first alternatives erratically a file r as a centroid of the primary cluster and provides the k-1 nearby facts to it, which will shape the primary cluster. Then the set of rules picks the file which has the biggest distance among itself and the primary centroid and sets it to the second one centroid. The ith centroid is created through with inside the identical way, primarily based totally on the space among the ith file and all of the happened centroids. Subsequently every step of centroid formation, the algorithm provides the k-1 closest facts to the centroid to shape the clusters. At the stop of this procedure, all of the clusters formed incorporate k facts, if there are ungrouped facts. The set of rules repeats the closing facts and enclosure every file into the nearest cluster, i.e., having the slightest space with its centroid.

An adaptive k-anonymity set of rules, known as AKA was proposed by Arava and Lingamgunta [14]. It is primarily centered totally on KOC's regular approach for locating the fine seed values. AKA begins off evolved with computing the variety of clusters p = no tuples/k value. For complete file in every institution, it computes k-closeness with each different file and types them in descending order. Then, it units in each institution the facts with minimal and most closeness as preliminary centroids (i.e., 2 * p seeds) and builds the clusters. The closing facts are allotted to their adjacent clusters, such that each cluster ought to have k cluster individuals. The extra facts (i.e., that have sizes special to k) are restructured and attached to their nearest clusters. For the clusters with scopes advanced to k, the procedure generates new clusters with insignificant of k facts. A present clustering set of rules can be implemented to the stay clusters, with sizes not as good as k, to distribute their facts.

Weighted k-member clustering set of rules known as (WKMCA) was proposed by Byun et al. [9]. The proposed set of rules is a changed k-individual to lessen the have an effect on outliers at the clustering effect. For this, WKMCA provides a biased level wherein a chain of weighting signs was assigned to assess the outlyingness of facts which will expedite filtering out the outliers. Thereby, k-individuals are primarily centered totally on the ones signs to acquire k-anonymity.

k-anonymity primarily based totally on nature-stimulated optimization procedures:

Lunacek et al. [15] projected a brand new crossover operator and carried out a Genetic Algorithm- primarily centered totally k-anonymity method with the suggested crossover operator to reveal the gain of the usage of the brand new operator over traditional crossover operators. Lin and Wei [16] projected a Genetic Algorithm (GA)-primarily centered totally clustering method for accomplishing k-anonymity. In this method, the preliminary populace of GA is fashioned primarily based totally on Hybrid Method anticipated [17]. A candidate answer of populace encoded through a chromosome and includes no rarer than k genes, wherein every gene suggests the index of a report with inside the authentic dataset. The set of rules makes use of most effective choice and crossover operations of GA. Mutation isn't always completed because of the set of rules makes use of the authentic report indexes which cannot be reformed.

Run et al. [18] proposed a hybrid seek technique primarily based totally on Tabu Search (TS) and Genetic Algorithm (GA) to acquire k-anonymity. In the projected technique, TS is embedded right into a traditional GA to carry out the position of mutation. Bhaladhare and Jinwala [19] anticipated a Fractional Calculus-primarily centered totally Bacterial Foraging Optimization Algorithm referred to as FC-BFO to generate a most reliable clustering. The goal of FC-BFO is to enhance the optimization cap potential and convergence velocity of BFO set of rules through making use of to it the idea of FC in its chemotaxis step. Effectively, the FC-BFO gives a higher records loss and execution time than BFO.

Wai et al. [20] proposed a huge statistics private maintenance method primarily based totally on Hierarchical Particle Swarm Optimization (HPSO). The suggested method is constructed upon MapReduce Hadoop groundwork to deal with the scalability problems of huge statistics. It includes stages; The first degree is HPSO clustering. The set of rules generates a MapReduce activity to provide the predefined amounts of intermediate clusters, characterized through particles, then a MapReduce activity of HPSO clustering is accomplished on every cluster through iteratively appearing Map and Reduce stages till the quantity of statistics individuals in every particle beat k. In the second one degree, the occasioned clusters are generalized to be converted into their anonymized paperwork. The Map step truly permits all statistics individuals of every intermediary cluster to its corresponding Reduce step which per-paperwork HPSO clustering activity to provide k-anonymized clusters.

Madan and Goswami proposed hybrid optimization algorithms to acquire k-anonymity referred to as Dragon-PSO [21] and GWO-CSO method [22]. The Dragon-PSO set of rules associations the Dragonfly Algorithm (DA) and Particle Swarm Optimization (PSO) through adapting the replace system of DA the usage of PSO. Madan and Goswami [23] projected an anonymity version for statistics issuing primarily centered totally on K-DDD degree, Dragonfly operators-primarily centered totally Genetic Algorithm referred to as Duplicate-Divergence-Different homes enabled Dragon Genetic (DDDG) set of rules. The head step, referred to as k-DDD anonymization, is the transformation of authentic database to k-DDD database primarily centered totally at the projected k-DDD degree. k-DDD degree transforms the authentic database through producing "k" quantity of identical facts, "k" quantity of Divergence in touchy attributes, and "k" quantity of Unlike provider companies in every cluster of the database. The subsequent phase is implemented D-Genetic set of rules on k-DDD database. D-Genetic Algorithm is shaped via the change of Genetic Algorithm (GA) with the Dragonfly Algorithm (DA).

k-anonymity in BigData MapReduce

Several privacy preservation algorithms fit into the MapReduce framework to perform parallel execution of large datasets. This is to address the scalability problem of BigData. MapReduce framework computes using map and reduces functions. Data from distributed file system is divided into number of chunks and assigned to map function and secretes a list of key/value pairs. In the subsequent section, Reducer syndicates the values fitting to every distinct key permitting to several functions and engraves the result to an output file. Thus the MapReduce function solves the scalability problem of BigData. LeFevre et al. [24] spoke the scalability problem of anonymization algorithms using scalable decision trees and sampling techniques. Fung et al. [25] projected the Top-Down Specialization methodology to yield anonymous datasets without data exploration problem. Ke et al. [26] proposed the Bottom-up generalization to anonymize large datasets. Yavuz et al. [27] proposed the data anonymization in large-scale dataset generated in real-time application using the Apache spark.

Proposed Approach

K-anonymization based on black hole algorithm in Big Data (KAB-BD) was proposed in the following section. The KAB-BD starts with initial populations of stars in each chuck of Map phase in MapReduce framework, clustering-centered k-anonymous solutions, and then progresses the population to discover the superlative k-anonymous solution, i.e., having the lowest information loss which is done in Reduce phase of the MapReduce framework. As the projected algorithm originates from black hole algorithm, the development of population to an optimal resolution is completed by moving all the stars to the best solution, characterized by the black hole.

Algorithm: k-anonymity-based black hole algorithm (KAB) in MapReduce

i. Initialize the map function with original population of stars
ii. In each chunk of Map phase, estimate the fitness of every star and customary the top star as the b-hole
iii. The fitness is updated and all stars are moved toward the b-hole
iv. If a star touches the finest location than the b-hole, it suits the b-hole and vice versa
v. If a star becomes too close to the b-hole, a new star is created to replace the old one and its fitness is evaluated
vi. Output from each chunk of Map is fed as input to Reduce phase and step (iii), (iv), and (v) is repeated
vii. If determined amount of iteration is reached, the algorithm stops

Step: 1 Generation of Initial Population

In the first step, on each map function, candidate solutions were selected based on clustering algorithm where solution signifies a k-anonymous clustering. The records in a cluster must be as alike as likely to acquire worthy data quality. This guarantees that fewer distortions are desirable to generalize the records from the identical cluster as a consequence subsequent of getting worthy data quality. *Normalized Certainty Penalty* (NCP) [28] is one such metric used to attain this objective and cost dimension of clustering algorithm. NCP is a proficient and easy to usage metric that deals the degree of information loss produced by the anonymization method.

Step: 2 Evaluation and Selection of the Black Hole

In the evaluation step followed by initialization, fitness function for every star is calculated to find the best star. The star with the finest fitness value is the black hole selected. After modifying the population, the fitness function of every star is weighed and the finest star, which has the finest fitness value, is selected as the black hole.

Step: 3 Update the Positions and the Fitness of the Stars

In the next subsequent step, totally the candidates move near the top candidate which is the black hole. This move can be done by shifting the location of each star.

Three possible scenarios are found in these situations:

i. After shifting the stars to new locations, star can reach the finest location than black holes with lesser fitness value. In this case, star befits the black hole and vice versa (interchange their locations and fitness).

ii. Star marks the event horizon of the black hole; in such a case the star will be absorbed by the black hole and switched by a new star.
iii. Neither of the two previous scenarios nor in this occasion is the locations and the fitness just rationalized. Once all the stars are relocated, subsequent reiteration precedes place with the novel positions of stars and black holes and their equivalent objective functions.
iv. Step. 2 and step. 3 is iterated in Reduce phase until the supreme quantity of iteration is seen and the algorithm terminates.

Experimental Result

In this section, we calculate the quality of our proposed algorithm. For preserving privacy, data utility and information loss are the main objectives of every anonymization algorithm. KAB is measured by evaluating these two objectives. Anonymized data utility is measured based on the different privacy levels characterized by k.

Evaluation Metrics and Experimental Setup:

Metric preferred for this is Classification Accuracy (CA), to calculate the rate of k-anonymous clusters in the anonymized data. A cluster is measured *correctly classified* if it fulfills k-anonymity criterion. To assess the data utility of our algorithm, we have used metrics namely *Classification Metric* (CM) and *Average Equivalence Class Size Metric* (CAVG) [24]. Information loss experienced by the unique data after anonymization process can be measured by two commonly used metrics: *Total Information Loss* (Total-IL) and *Normalized Certainty Penalty* (NCP). Total Information Loss is defined as the loss of accuracy when take a broad view of specific attributes. NCP *measures* the classification errors by penalizing equivalence classes that contain rows with different class labels.

The carrying out tests were accomplished on a Desktop PC with Intel Core 2.10 GHz CPU and 4 GB of RAM under Ubuntu operating system. The executions were built and run with Hadoop. The adult dataset was taken from the UCI machine learning repository for anonymization. The dataset comprises census data has a total of 32,561 instances underneath 15 attributes. Each record has the personal information and personal income-related information. Preprocessing is done at initial stage for the removal of duplicate and missing values. In adult dataset, age and education are the numerical attributes and remaining attributes are the categorical attributes. Totally, dataset has 30,162 records.

Results and Discussion

Data Utility with different levels of privacy is done by considering three different algorithms (k-anonymization, KAB-based k-anonymization, and BHA-based k-anonymization) and the experimental results are shown in Figs. 4.1, 4.2, 4.3, and 4.4 Data Utility with respect to information loss is increasing with the increasing value of k represented by *GenTotal-IL* and *GCP*, correspondingly for all the three algorithms is depicted in Figs. 4.1 and 4.2.

K represents the number of records in a cluster. If the value of k is large, extra records are found in single cluster. It is found that, the KAB-based k-anonymization presents the minimum information loss with respect to privacy level. BHA-based

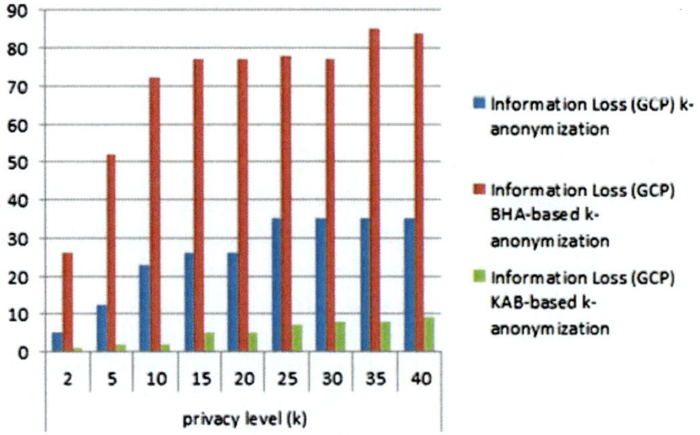

Fig. 4.1 Information loss (GCP) versus privacy level

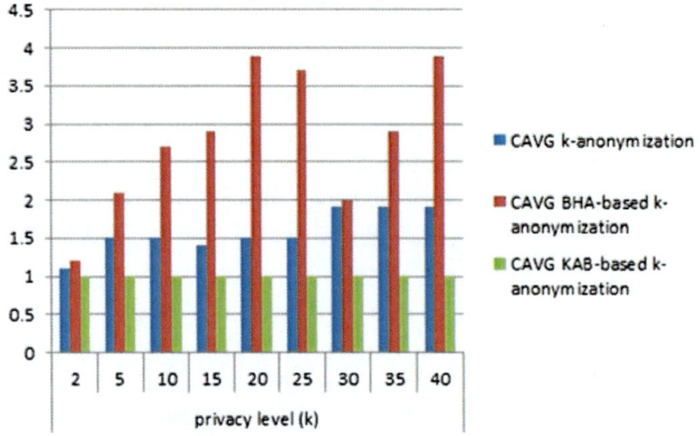

Fig. 4.2 *C*AVG versus privacy level

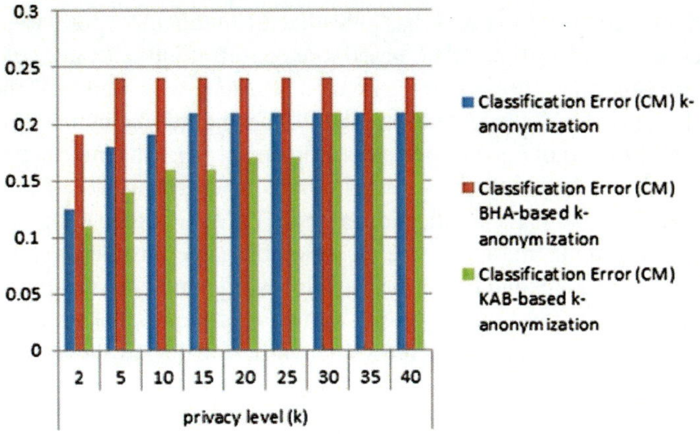

Fig. 4.3 Classification error (CM) versus privacy level

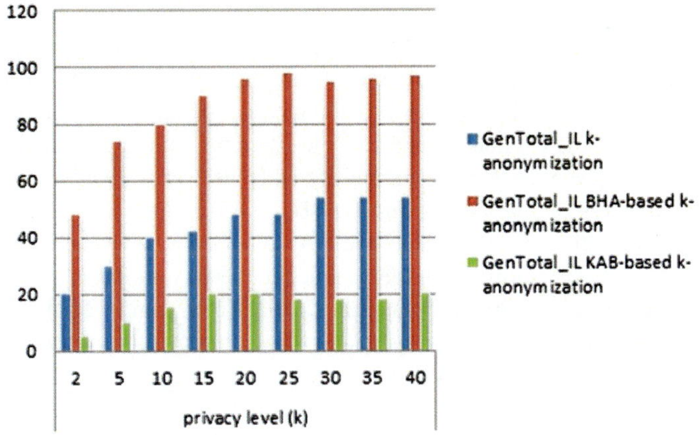

Fig. 4.4 Generalization Total Information loss (GCP) versus privacy level

k-anonymization has the poorest information loss since it doesn't follow any reliable metrics to organize the records in the cluster and the records are placed randomly. Also, BHA-based k-anonymization has the small convergence rate and requires a more quantity of iteration to cover.

It is found that KAB algorithm is an enhancement of BHA. KAB declines the distance among preliminary results and optimal results to accelerate the convergence rate BHA to create an optimal solution. Figure 4.3 considers the data utility of the entire three algorithms with respect to *C*AVG metric as the value of k rises. It is the reflection of the information lost from previous figures. Based on the observation, it is found that Mondrian Multidimensional forms equivalence classes, of sizes adjacent to ideal case which clarifies the reasonable information loss familiarized by k-anonymization. KAB algorithm generates correspondence classes of ideal sizes, i.e., equal to 1, which has subsidized to reducing information loss of KAB-based k-anonymization. BHA-based k-anonymization generates equivalence classes of variable sizes, because the amount of clusters, which governs the size of the equivalence classes, is calculated randomly; the slighter the number of clusters is, the greater the sizes of equivalence classes are, and higher the information loss it.

Figure 4.4 reports data utility, of the algorithms, with respect to *CM* as the value of k increases. It shows that classification errors of the KAB-based k-anonymization and k-anonymization increase with the increase of k-value. Instinctively, the higher the class dimensions are, and better the probability of result classification errors is. Generally, CM presented by BHA-based k-anonymization is the same. From the figure, it is detected that KAB-centered k-anonymization declares less classification errors than the other algorithms.

Conclusion

K-anonymization centered on BHA (KAB) is proposed in this work. The main objective is to find the optimal clustering-based k-anonymity in Map Reduce framework. This starts with a population of clustering-centered k-anonymous candidate solutions, on which BHA is applied. To measure the efficiency, our approach is compared with k-anonymity, BHA-centered k-anonymity, and clustering-centered k-anonymity techniques, in terms of data utility and scalability. To reduce the execution time, the above said algorithm is implemented in MapReduce framework. The simulation results report that KAB algorithm in MapReduce outperforms all the compared techniques in terms of data utility and scalability. Data Utility can be further enhanced by increasing the number of iterations and/or stars. In our future work, this can be improved by implementing in machine learning based approach.

References

1. Ciriani, V., di Vimercati, S.D.C., Foresti, S., Samarati, P., Yu, T.: k-anonymity. In: Jajodia, S., Yu T. (eds.) Security in Decentralized Data Management. Springer (2006)
2. Sweeney, L.: Datafly: A system for providing anonymity in medical data. In: Database Security XI, pp. 356–381. Springer, Boston, MA (1998). https://doi.org/10.1007/978-0-387-35285-5_22
3. Sweeney, L.: k-anonymity: a model for protecting privacy. Int. J. Uncertainty Fuzziness Knowl. Based Syst. **10**(05), 557–570 (2002). https://doi.org/10.1142/S0218488502001648
4. De Montjoye, Y.A., Radaelli, L., Singh, V.K.: Unique in the shopping mall: On the reidentifiability of credit card metadata. Science **347**(6221), 536–539 (2015). https://doi.org/10.1126/science.1256297
5. Meyerson, A., Williams, R.: On the complexity of optimal k-anonymity. In: Proceedings of the Twenty-Third ACM SIGMOD-SIGACT-SIGART Symposium on Principles of database systems, pp. 223–228 (2004). https://doi.org/10.1145/1055558.1055591. Moon, B., Jagadish, H.V., Faloutsos, C., Saltz, J.H.: Analysis of the clustering properties of the hilbert space-filling curve. IEEE Trans. Knowl. Data Eng. **13**(1), 124–141 (2001). https://doi.org/10.1109/69.908985

6. Hatamlou, A.: Black hole: A new heuristic optimization approach for data clustering. Inf. Sci. **222**, 175–184 (2013). https://doi.org/10.1016/j.ins.2012.08.023
7. Li, J., Wong, R.C.W., Fu, A.W.C., Pei, J.: Achieving k-anonymity by clustering in attribute hierarchical structures. In: International Conference on Data Ware-housing and Knowledge Discovery, pp. 405–416. Springer, Berlin, Heidelberg (2006). https://doi.org/10.1007/11823728_39
8. Chiu, C.C., Tsai, C.Y.: A k-anonymity clustering method for effective data privacy preservation. In: International Conference on Advanced Data Mining and Applications, pp. 89–99. Springer, Berlin, Heidelberg (2007). https://doi.org/10.1007/978-3-540-73871-8_10
9. Byun, J.W., Kamra, A., Bertino, E., Li, N.: Efficient k-anonymization using clustering techniques. In: International Conference on Database Systems for Advanced Applications, pp. 188–200. Springer, Berlin, Heidelberg (2007). https://doi.org/10.1007/978-3-540-71703-4_18
10. Loukides, G., Shao, J.: Capturing data usefulness and privacy protection in k-anonymization. In: Proceedings of the 2007 ACM symposium on Applied computing, pp. 370–374 (2007). https://doi.org/10.1145/1244002.1244091
11. Lin, J.L., Wei, M.C.: An efficient clustering method for k-anonymization. In: Proceedings of the 2008 International Workshop on Privacy and Anonymity in Information Society, pp. 46–50 (2008). https://doi.org/10.1145/1379287.1379297
12. Ni, S., Xie, M., Qian, Q.: Clustering based K-anonymity algorithm for privacy preservation. IJ Netw. Secur. **19**(6), 1062–1071 (2017)
13. Zheng, W., Wang, Z., Lv, T., Ma, Y., Jia, C.: K-anonymity algorithm based on improved clustering. In: International Conference on Algorithms and Architectures for Parallel Processing, pp. 462–476. Springer, Cham (2018). https://doi.org/10.1007/978-3-030-05054-2_36
14. Arava, K., Lingamgunta, S.:Adaptive k-anonymity approach for privacy preserving in cloud.Arab. J. Sci. Eng.,1–8(2019).https://doi.org/10.1007/s13369-019-03999-0
15. Lunacek, M., Whitley, D., Ray, I.: A crossover operator for the k-anonymity problem. In: Proceedings of the 8th Annual Conference on Genetic and Evolutionary Computation, pp. 1713–1720 (2006). https://doi.org/10.1145/1143997.1144277
16. Lin, J.L., Wei, M.C.: Genetic algorithm-based clustering approach for k-anonymization. Expert Syst. Appl. **36**(6), 9784–9792 (2009). https://doi.org/10.1016/j.eswa.2009.02.009
17. Lin, J.L., Wei, M.C., Li, C.W., Hsieh, K.C.: A hybrid method for k- anonymization. In: 2008 IEEE Asia-Pacific Services Computing Conference, pp. 385–390. IEEE (2008). https://doi.org/10.1109/APSCC.2008.65
18. Run, C., Kim, H.J., Lee, D.H., Kim, C.G., Kim, K.J.: Protecting privacy using k-anonymity with a hybrid search scheme. Int. J. Comput. Commun. Eng. **1**(2), 155 (2012)
19. Bhaladhare, P.R., Jinwala, D.C.: Novel approaches for privacy preserving data mining in k-anonymity model. J. Inf. Sci. Eng. **32**(1), 63–78 (2016)
20. Wai, E.N.C., Win, A.T., Tsai, P.W., Pan, J.S.: Privacy preservation in big data by particle swarm optimization. University of Computer Studies (Taunggyi) (2017)
21. Madan, S., Goswami, P.: A privacy preserving scheme for big data publishing in the cloud using k-anonymization and hybridized optimization algorithm. In: 2018 International Conference on Circuits and Systems in Digital Enterprise Technology (ICCSDET), pp. 1–7. IEEE (2018). https://doi.org/10.1109/ICCSDET.2018.8821140
22. Madan, S., Goswami, P.: A novel technique for privacy preservation using k-anonymization and nature inspired optimization algorithms. In: Proceedings of International Conference on Sustainable Computing in Science, Technology and Management (SUSCOM), Amity University Rajasthan, Jaipur, India (2019). https://doi.org/10.2139/ssrn.3357276
23. Madan, S., Goswami, P.: k-DDD measure and mapreduce based anonymity model for secured privacy-preserving big data publishing. Int. J. Uncertainty Fuzziness Knowl. Based Syst. **27**(02), 177–199 (2019). https://doi.org/10.1142/S0218488519500089
24. LeFevre, K., DeWitt, D.J., Ramakrishnan, R.: Incognito: Efficient full-domain k-anonymity. In: Proceedings of the 2005 ACM SIGMOD International Conference on Management of Data, pp. 49–60 (2005). https://doi.org/10.1145/1066157.1066164

25. Fung, B.C.M, Wang, K., Yu, P.S.: Anonymizing classification data for privacy preservation. IEEE Trans. Knowl. Data Eng. **19**(5), 711–725 (2007)
26. Ke, W., Yu, P.S., Chakraborty, S.: Bottom-up generalization: A data mining solution to privacy protection. In: Proceedings of 4th IEEE International Conference on Data Mining, ICDM'04, pp. 249–256 (2004)
27. Yavuz, C., Seref, S.: BigData anonymization with spark. In: IEEE International Conference on Computer Science and Engineering (UBMK) (2017)
28. Xu, J., Wang, W., Pei, J., Wang, X., Shi, B., Fu, A.W.C.: Utility-based anonymization for privacy preservation with less information loss. ACM SIGKDD Explor. Newsl. **8**(2), 21–30 (2006). https://doi.org/10.1145/1233321.1233324
29. Pramanik, M.I., Lau, R.Y., Zhang, W.: K-anonymity through the enhanced clustering method. In: 2016 IEEE 13th International Conference on e-Business Engineering (ICEBE), pp. 85–91. IEEE (2016). https://doi.org/10.1109/ICEBE.2016.024

Chapter 5
Sleep Quality Analysis Using Motion Signals and Heart Rate

R. Vijayalakshmi, Prakash Rajiah, A. Lakshmi Sangeetha, and A. Balaji Ganesh

Introduction

Sleep can be considered as a divine gift for humanity. It is a period of rejuvenation and repair. Sleep is very important for maintaining physical and mental well-being. With the growing technology and demands of daily life, getting adequate sleep has taken a back seat. Another factor is the sleep-related disorders that come as unwelcomed guests to many. Whichever may be the cause, it is important to monitor our sleep pattern and make changes if it is disturbed. Sleep quality monitoring has gained importance over the years and many methods, either in a clinical environment or at home have emerged extensively.

Polysomnography is a test used to detect sleep disorders. It is designed in a way to record brain waves, oxygen level in the blood, heart rate, breathing, and movements of eyes and legs. It is carried out at a sleep disorder unit in a hospital or at a sleep center. Patients are attached to numerous wires and sleeping in an unfamiliar environment

R. Vijayalakshmi (✉)
Electronic System Design Laboratory, Velammal Engineering College, Chennai, Tamil Nadu, India
e-mail: viji92@hotmail.com

P. Rajiah
Emsensing Technologies Ltd, Chennai, Tamil Nadu, India

A. Lakshmi Sangeetha
Department of Electronics and Instrumentation, Velammal Engineering College, Chennai, Tamil Nadu, India

A. Balaji Ganesh
Electronic System Design Laboratory, Velammal Engineering College, Chennai, Tamil Nadu, India

Ch. Satyanarayana et al. (eds.), *Proceedings of the International Conference on Computer Vision, High Performance Computing, Smart Devices and Networks*, Advanced Technologies and Societal Change,
https://doi.org/10.1007/978-981-19-4044-6_5

makes it difficult for them. Hence, people look out for more patient-friendly methods of sleep analysis.

An affordable unobtrusive sensing environment with a bed accelerometer and passive infrared (PIR) motion sensors in all rooms is explained in [1]. This follows the general framework of Internet of Health Things for monitoring the elderly's sleep patterns and assessing their sleep quality. The sleep apnea detection system in [2] used Spo2 and ECG signals and the efficiency of various machine learning algorithms detecting apnea was compared. Also, the importance of feature selection was emphasized.

An automatic sleep scoring algorithm was developed in [3] to analyze ECG signal to estimate sleep quality and sleep apnea using Convolution neural network. This algorithm was non-invasive and could be done at home. A smart bed was created in [4] which gathered motion, heart rate, and breathing data of the occupant and was used to monitor sleep and assess REM and non-REM sleep. Fallmann and Chen [5] dealt with importance of sleep and the need for monitoring sleep patterns. Recent trends in sleep monitoring and the various technologies associated with it were also discussed.

Authors in [6] compared sleep parameters generated by a PAM with the polysomnography (PSG) and it was found to be correlated. An algorithm to estimate the quality of sleep, by analyzing the cyclic alternating pattern rate was developed in [7]. The algorithm used a single-lead electrocardiogram to produce a spectrographic measure of the cardiopulmonary coupling which was fed to a classifier to estimate the non-REM sleep and the presence of the cyclic alternating pattern.

The importance of sleep monitoring in elderly persons is discussed in [8]. The viability of cardiorespiratory and physical measurements over polysomnography is discussed. The issues related to the new sleep monitoring methods are also discussed. An actigraphy recording device was developed in [9] in order to avoid the polysomnography method. Sleep–wake moments were detected using concepts of movement density evaluation and adaptive windowing. Berke Erdaş et al. [10] explores the different features which can be extracted from the accelerometer sensors attached to a patient's body for activity recognition. From this feature set, the most relevant features are selected using feature selection.

This paper is aimed at predicting the quality of sleep and how different factors affect sleep in adults. It incorporates machine learning algorithms for sleep–wake classification and hence estimates sleep efficiency. It uses actigraphy, R-R interval values, inclinometer data, and heart rate data for determining details about sleep.

Materials and Methods

Actigraphy, i.e., accelerometer data contain valuable information about the body position of a person. When a person is sleeping properly their body movement is negligible, when they have disturbed sleep or when they are awake, their body position changes constantly. Using this, we can use the body position data as a marker of good sleep.

When a person is asleep the heart rate tends to be lower than the heart rate when the person is awake or active. Inclinometer is a device used to measure the body position/inclination with respect to gravity. This also provides useful information about the sleeping habits and postures. RR interval denotes the time elapsed between two successive R-waves of the QRS signal on the electrocardiogram which also changes during sleep and wake periods. Accelerometer, inclinometer data, heart rate, and R-R interval information are used for sleep quality analysis in this paper. The sleep quality analysis procedure is a series of steps as depicted in Fig. 5.1.

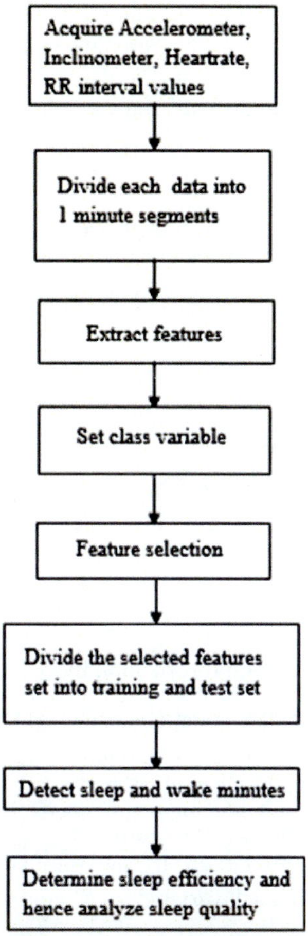

Fig. 5.1 Flowchart for sleep quality analysis

Data Used

The data for this paper is taken from the Physionet repository Multilevel Monitoring of Activity and Sleep in Healthy people (MMASH) dataset [11]. Overnight accelerometer, inclinometer, heart rate, and R-R interval values of 20 volunteers were taken and divided into 1 min epochs. They are of age group 20–40 years.

Feature Extraction

Feature extraction is one of the most important steps of signal analysis. The signal is approximated into a feature set which is given to machine learning algorithms. The following features are extracted for each input discussed above: Mean, Median, Variance, Standard Deviation, Maximum, Minimum, Entropy, Skew, Kurtosis, Interquartile range, and Mean Absolute Deviation.

Feature Selection

Not every feature of the signal contributes to classification in a positive way. Hence, the need for determining the set of features that are dominant and assist the classification arises. The class variables are assigned with values and given to feature selection algorithms. Here five algorithms are used and the dominant features are selected:

Recursive feature elimination: This algorithm recursively removes the attributes that doesn't contribute and build a model on those attributes which remain. It ranks the feature based on their importance and accuracy.

Random forest importance: Random Forests is a type of a Bagging Algorithm that aggregates a specified number of decision trees. The tree-based strategies used by random forests rank by how well they improve the purity of the node.

Extra trees classifier: This classifier implements a meta estimator which fits a number of randomized decision trees on various sub-samples of the dataset and makes use of averaging to improve the predictive accuracy and control over-fitting.

Information gain: This calculates the reduction in entropy from the transformation of a dataset. It can be used for feature selection by evaluating the Information gain of each variable in the context of the target variable.

Variance threshold method: It is a baseline approach that removes all features where variance doesn't meet some threshold. it removes all zero-variance features.

Classification

The dominant features of the signals are given to the classifiers. The performance is compared with actual features. The feature set was split into training (70%) and test set (30%). This was given to classifiers and the performance metrics were compared:

AdaBoost classifier: It classifies by fitting a classifier on the original dataset and then the additional copies of the classifier are fitted on the same dataset with an exception that the weights of incorrectly classified instances are adjusted.

K-nearest neighbor classifier: In this, k closest neighbors are taken and analyzed where most of these neighbors are classified. The new item is classified in that.

Logistic Regression: This technique is used to predict the probability of a categorical dependent variable. In logistic regression, the dependent variable is a binary variable that contains data coded as 1 (yes) or 0 (no), the model predicts Prob($Y = 1$) as a function of X.

Random forests: This algorithm creates decision-making trees on randomly selected data samples, gets prediction from each tree, and selects the best solution by majority vote.

Sleep Efficiency and Sleep Quality

With the growing technology and demands of daily life, getting adequate sleep has taken a back seat. Another factor is the sleep-related disorders that come as unwelcomed guests to many. Whichever may be the cause, it is important to monitor our sleep pattern and make changes if it is disturbed. Sleep quality monitoring has gained importance over the years and many methods, either in a clinical environment or at home have emerged extensively.

Sleep efficiency is the relation of time that a person sleeps to the amount of time a person spends in bed. This percentage is calculated by fraction of Total Sleep Time by Total Time in bed. The Pittsburgh Sleep Quality Index (PSQI) provides a subjective sleep quality measure and patterns. The PSQI is calculated using the inputs given by volunteers for questionnaires. Using the calculated sleep efficiency, the accuracy of the PSQI can be checked. Good sleep is indicated by a PSQI score of 5 and below.

Results and Discussion

The data given as input to the machine earning algorithms was the sensor data of volunteers starting from the period they settled down in bed until they got out of bed after sleep. So, the data contains wake period (awake in bed) and sleep period. The

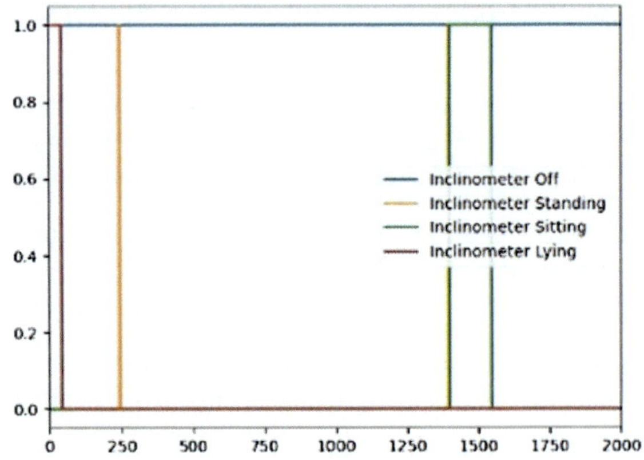

Fig. 5.2 Inclinometer values during sleep

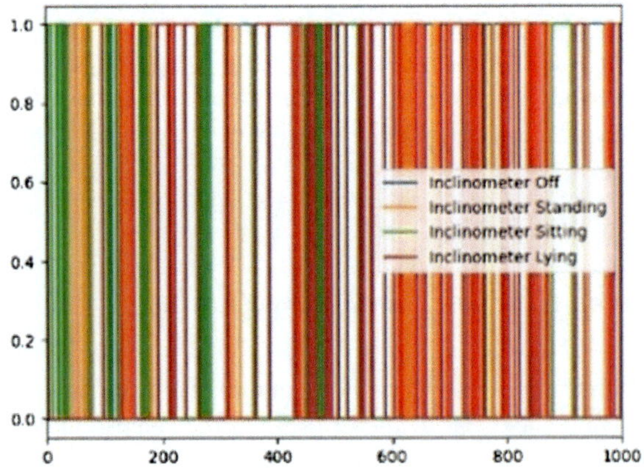

Fig. 5.3 Inclinometer values when awake

various inclinometer values during sleep and awake period are depicted in Figs. 5.2 and 5.3.

The heart rate and R-R interval values during sleep and awake state are shown in Figs. 5.4 and 5.5, respectively.

The accelerometer values during sleep and wake stages have a large difference and can be seen in Figs. 5.6 and 5.7.

Using the data given, the algorithms used were able to detect the sleep and wake periods efficiently and the results are given in Table 5.1. It can be seen that feature election has improved the results and KNN and Random Forest algorithms have performed better than the other two.

Using the sleep, wake results of the classification algorithms total time in bed and total sleep time are calculated. Using this, sleep efficiency is calculated and compared with the efficiency given in the database. It can be seen that the calculated sleep efficiency matches with the given sleep efficiency. Volunteers with a PSQI score of 6 and above have poor sleep efficiency, and those with PSQI score of 5 and below have improved sleep efficiency of above 85% (Table 5.2).

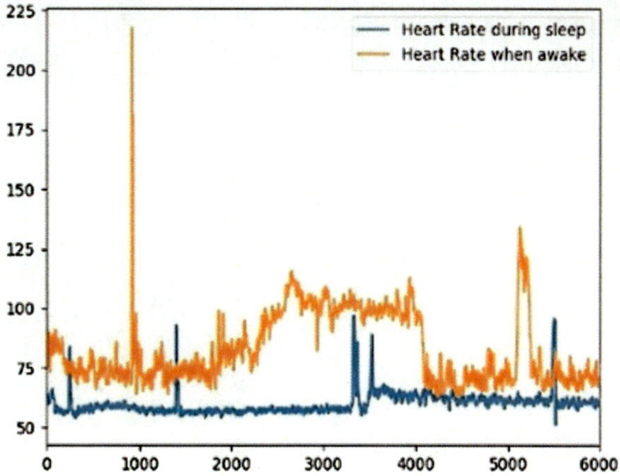

Fig. 5.4 Heartrate during sleep and awake

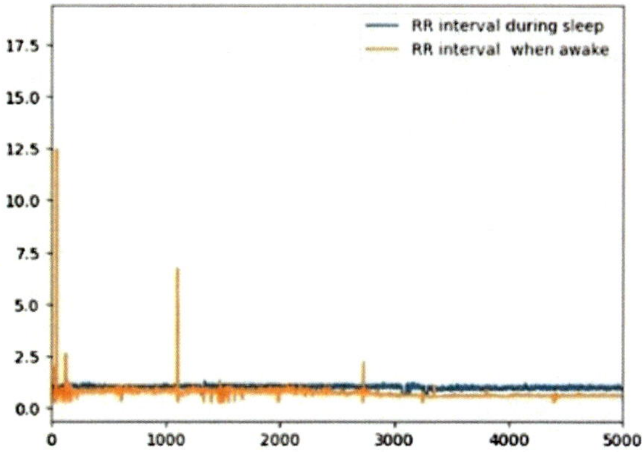

Fig. 5.5 R-R during sleep and awake

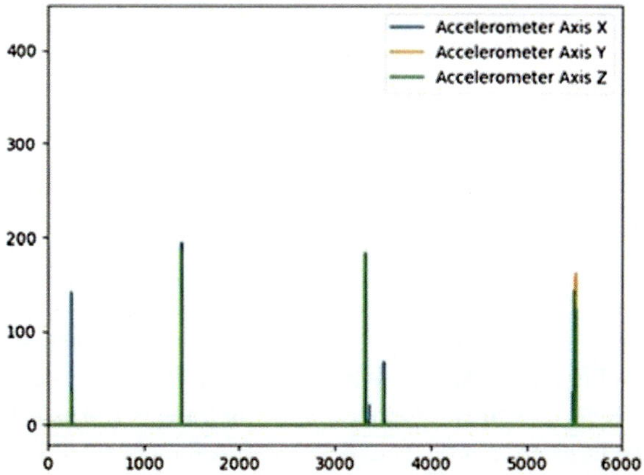

Fig. 5.6 Accelerometer values during sleep

Fig. 5.7 Accelerometer values when awake

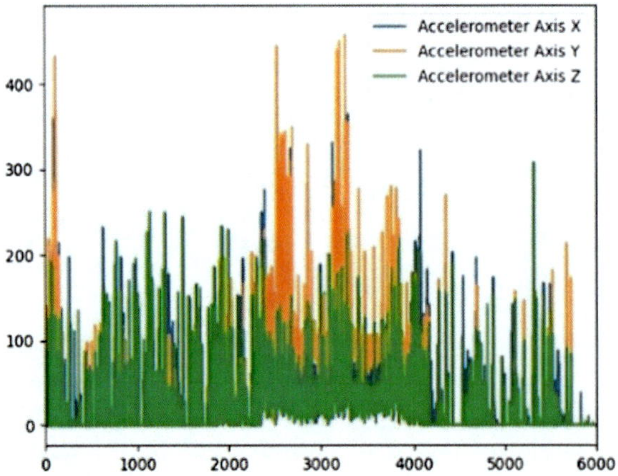

Table 5.1 Classification results

Without feature selection				With feature selection			
AdaBoost	KNN	Linear regression	Random forest	AdaBoost	KNN	Linear regression	Random forest
93.92	95.62	93.59	94.87	95.77	97.43	95.83	96.76

Table 5.2 Comparing calculated and given sleep efficiency

Volunteer	Calculated sleep efficiency	Given sleep efficiency	PSQI
1	85.42	87.27	5
2	76.61	79.23	8
3	82.88	84.3	9
4	72.9	75.33	7
5	88.46	90.78	2
6	85.11	86.93	3

Conclusion

From this study, it can be seen that sleep quality analysis can be done easily without the hassles of overnight polysomnography, by using accelerometer, heart rate, and inclinometer sensors. We experimented with various classification algorithms out of which K-nearest neighbor algorithm has performed better. From the results, it can be inferred that the calculated sleep efficiency used in this method matches with the PSQI scores and the actual efficiency thus validating this method to be used as a sleep quality analysis tool.

Acknowledgements The authors wish to acknowledge Department of Science and Technology (DST), Government of India, for sanctioning the project "Wireless body sensor network supported Sleep Disorder Analysis system", DST No. SEED/TIDE/2018/20/G, from which tools are required for this paper are used.

References

1. Kim, J.-Y., Chu, C.-H., Kang, M.-S.: IoT-based unobtrusive sensing for sleep quality monitoring and assessment. IEEE Sens. J. **21**(3), 3799–3809 (2021)
2. Ramasubbu, S. et al.: Analysis of sleep apnea considering biosignals from peripheral capillary oxygen saturation level and electrocardiogram data. In: Pandian, A.P., Palanisamy, R., Ntalianis, K. (eds.) Proceedings of International Conference on Intelligent Computing, Information and Control Systems. Advances in Intelligent Systems and Computing, vol. 1272. Springer, Singapore (2021)
3. Mendonça, F., Mostafa, S.S., et al.: A method for sleep quality analysis based on CNN ensemble with implementation in a portable wireless device. IEEE Access **8** (September 2020)
4. Laurino, M., Arcarisi, L., et al.: A smart bed for non-obtrusive sleep analysis in real world context. IEEE Access (March 2020)
5. Fallmann, S., Chen, L.: Computational sleep behavior analysis: a survey. IEEE Access (October 2019)
6. Spielmanns, M., Bost, D., Windisch, W., Alter, P., Greulich, T., Nell, C., Storre, J.H., Koczulla, A.R., Boeselt, T.: Measuring sleep quality and efficiency with an activity monitoring device in comparison to polysomnography. J. Clin. Med. Res. **11**(12), 825–833 (2019)
7. Mendonça, F., Mostafa, S.S., Morgado-Dias, F., Ravelo-García, A.G.: Sleep quality estimation by cardiopulmonary coupling analysis. IEEE Trans. Neural Syst. Rehabil. Eng. **26**(12), 2233–2239 (2018)
8. Matar, G., Lina, J., Carrier, J., Kaddoum, G.: Unobtrusive sleep monitoring using cardiac, breathing and movements activities: an exhaustive review. IEEE Access **6**, 45129–45152 (2018)
9. Kuo, C., Liu, Y., Chang, D., Young, C., Shaw, F., Liang, S.: Development and evaluation of a wearable device for sleep quality assessment. IEEE Trans. Biomed. Eng. **64**(7), 1547–1557 (July 2017)
10. Berke Erdaş, Ç., Atasoy, I., Açıcı, K., Oğul, H.: Integrating features for accelerometer-based activity recognition. Procedia Comput. Sci. **98,** 522–527 (2016)
11. Rossi, A., Da Pozzo, E., Menicagli, D., Tremolanti, C., Priami, C., Sirbu, A., Clifton, D., Martini, C., Morelli, D.: Multilevel monitoring of activity and sleep in healthy people (version1.0.0). PhysioNet (2020). https://doi.org/10.13026/cerq-fc86

Chapter 6
Prediction and Performance Assessment in Woman Handball Athletes by Employing Machine Learning Methods

Kolla Vivek, J. Harikiran, Kunda Suresh Babu, and Ch. Anil Carie

Introduction

Handball is recognized as a game needing endurance, agility, endurance, and an intermittent pace, with sporadic features of the game containing rapid defense and attack [1]. In addition, handball is performance-oriented, i.e., winning a game depends on performance of the athletic and incorporates items that are mechanical, strategic, and psychological.

As a consequence of enforcing modish regulations, the abilities required and relevant to players have increased considerably in the past few years. Handball seems to have become a quick and intense sport in which players are supposed to do better with a disciplined run, drive, leap, shoot, move, and block capabilities. Consequently, the information from multiple workout experiments for the evaluation of actual game expertise in the field is very essential to examine. Sport performance analysis and methods permit sports managers, selection committees, and trainers to evaluate player's capability objectively. Thus, the capability and potential of the players have become a key factor of selection and training. Researchers have implemented several

K. Vivek (✉)
Computer Science and Engineering, QIS College of Engineering and Technology, Ongole 523001, India
e-mail: kollavivek@gmail.com

J. Harikiran · K. Suresh Babu · Ch. Anil Carie
School of Computer Science and Engineering, VIT-AP University, Amaravati, India
e-mail: harikiran.j@vitap.ac.in

K. Suresh Babu
e-mail: suresh.20phd7019@vitap.ac.in

Ch. Anil Carie
e-mail: anil.carie@vitap.ac.in

Ch. Satyanarayana et al. (eds.), *Proceedings of the International Conference on Computer Vision, High Performance Computing, Smart Devices and Networks*, Advanced Technologies and Societal Change,
https://doi.org/10.1007/978-981-19-4044-6_6

techniques and approaches to assess the most critical attributes of sport and exercise biomechanics, the science of life, and psychobiology [2]. Because of the low error rate and high accuracy rate, artificial intelligence (AI) and its techniques are widely used to develop robust solutions to complex problems. Deep learning as a subgroup of ML and ML which is a field of AI is also used in different fields to develop solutions to complex problems. ML always utilizes experience to predict the results, and the prediction rate always depends on the attributes and features of the dataset. A very little research was done on handball sport using ML techniques. The remaining parts of the work are presented as follows. Section 2 discussed related studies. It is in Sect. 3 that the proposed method is presented. The findings were addressed in detail in Sect. 4. Section 5 concludes with findings and recommendations for the future.

Related Works

Many experts and practitioners developed prediction models using machine learning for analyzing the player's performance in different sports sciences and games like volleyball, basketball, handball, cricket, swimming, football, etc. Jovanovic et al. [1] conducted experiments on 20 female respondents aged between 16 and 25 years. The results had proved an R-squared value of 0.876. Tests conducted on players included SAMO reaction-agility test and the Illinois test. Gomez-Lopez et al. [2] considered the expectation capacity over the convictions that causes the achievement in handball sport. The example was included 444 elite competitors (233 young men and 211 young ladies. The relation between self-regulation techniques, mindfulness practice, and success was discussed by Popa et al. [3]. The study is comprised of 288 handball representatives from Romania. The participants were 30% male and 70% female, varying in age from 12.01 to 14 years. In a hierarchical multiple regression, the parameters (state consciousness of the individual, self-monitoring, and self-efficacy) clarified 87% of the variation in sports success.

Hermassi et al. [4] conducted experiments on 72 young handball players whose age group lies between 15.2 and 16.4 years, and performance measurements of lower limbs in young handball players were taken into consideration. They assessed the player's performance twice, on separate days. Tests like squat jumps, counter-movement jumps, 5m sprint, 10-m sprint, and handball sport skill test were conducted and recorded finishing times using electronic timing gates. The model forecasted the player's performance with R-squared values between 0.52 and 0.68.

Sekulic et al. [5] conducted experiments on 32 male futsal players whose age group lies between 26.22 and 31.22 years, body height lies between 182.13 and 187.99 cm, and body mass lies between 77.43 and 85.00 kg. Tests like reactive agility (RAG), change of direction speed (CODS), and 10 m sprint were conducted and recorded finishing times of players using power timer 300. Soslu et al. [6] conducted experiments on 23 male basketball players whose age group lies between 23.2 and 26.7 years, body height lies between 197.1 and 206.1, and body mass lies between 95.3 and 105.3 kg; anthropometric performance measurements, vertical jump measurements, sprint performance evaluation, anaerobic performance measurements of isokinetic knee strength, muscle strength in basketball players were taken into consideration. The investigation was led over 1 week, during which the players did not take an interest in some other preparation or matches. Tests like Wingate anaerobic test, T-drill were conducted and recorded finishing times.

All of the above-mentioned experiments are intended to execute ML models to establish standards of player exhibitions of particular skills in different areas of the game, as well as to help mentors decide on reasonable decisions in terms of squad or individual player determination [7–10]. Only a certain number of research on the use of AI models in sports science have been published. In addition, each game has a remarkable structure, physical abilities, and prerequisites. For certain activities, games, and sports to be settled, the diverse and dynamic layout of athletic exhibitions requires an appropriate ML model [11–14]. As far as we could know, this is the main investigation to utilize ML models to anticipate explicit exhibitions in handball players. Further investigations may be required for various games, age gatherings, and genders.

Four ML models are used in this article to predict the performances of women handball players and allow mentors to accurately assess the performance of players before games. Furthermore, huge elements impacting the thought about exhibition abilities were resolved to improve the player execution. The intension and considerations of this analysis can be defined as follows, because of the previously mentioned data and a literature survey: To actualize a few ML techniques to determine the optimal standard for the competencies considered and to carry out a related test with different measurements.

(i) To consider numerous ability skills of players for the prognosis.
(ii) To anticipate player's ability in four competencies and to help mentors with the effective choice of players in games.
(iii) To decide the elements that influence the considered competencies and help coaches to concentrate on critical variables to improve player performance of specific competencies.

Proposed Methodology

Dataset

Data related to 52 players whose age group lies between 20.2 and 26.1 years, height lies between 164.1 and 170.7 cm, bodyweight lies between 62.4 and 72.1 kg, and body mass index (BMI) lies between 23.4 and 25.7 kg m^2 were collected from Kaggle repository. For two intervals, socioeconomic data and psychometrics were collected. For the ML models, a total of 23 attributes (Table 6.1) and attribute metrics were captured, and 117 occurrences were captured for training samples. Each metric was captured for every player in parallel during two separate intervals, and a dataset

Table 6.1 Dataset features

Age (years)	Backside middle upper arm double skin density (ST) (mm)
Supraspinal double skin density in millimeters	Wingate average power (AP) in kilograms
Athletic's body weight in kilograms	Front side middle upper arm double skin density (mm)
Midriff double skin density in millimeters	Wingate comparative mean power (cMP) (kg W^{-1})
Athletic's body length in meters	Medial calf double skin density in millimeters
Thigh double skin density in millimeters	Wingate peak power (PP) in kilograms
Quetelet index (kg/m^2)	lower shoulder blade double skin density (mm)
Midaxillary double skin density (mm)	Wingate relative peak power (rPP) (kg W^{-1})
Ten-meter sprint race (SP10) in seconds	Squat jump (SJ) (cm)
Thorax double skin density (mm)	Handball sport-*skill* test (HSST) in seconds
Twenty-meter sprint race(SP20) in seconds	Squat jump on roes (SJT) in centimeters
Shuttle run (SR) in kilometers per hour	–

involving two or more variable quantities was created. Any time-related information was not recorded in the dataset. Nine skin fold points have been chosen (upper arm middle front, upper arm middle back, lower shoulder blade, supraspinal, thorax, midriff, lineaaxillaris, upper leg, and anterior calf), and these points have been quantified in compliance with the recommendations in [15]. Quetelet index (QI) can be formulated by taking measurements of player's weight and height into consideration. It is expressed in Eq. (6.1) by its formula.

$$\text{QI (or) BMI} = \frac{m}{h^2} \tag{6.1}$$

where 'm' is the weight of the players measured in kilograms (kg), and 'h' is the height of the players measured in meters (m).

Using the Wingate test, Bar-Or [16] registered normal force, relative normal force, top force, and relative pinnacle strength. On a 20 m straight track, the speed was measured, and information was registered at 10 and 20 m. For any athlete, experiments were rehashed numerous times for time intervals of minutes. For the test times of 10 m (SP10) and 20 m (SP20), the fastest time was recorded [17]. Following the guidelines of Iacono et al. [17], the handball sport-skill test (HSST) was conducted. For each athlete with a 60s gap between tests, squat jumps (SJ) and squat jumps on toes (SJT) measurements were carried out on a piezoelectric force plate 3 times, and the best jump was recorded as per Moir [18].

Machine Learning Models

The four simplest standard ML models considered in this research for the expectation and investigation of the abilities of players are SLR, CT, SVR, and RBFNN.

Simple Linear Regression (SLR)

One of the critical and fundamental models of ML for prognostication is SLR. It is mainly used for the prognostication of information that generally has a linear association between its features and occurrences. For a 'Z' tagged dataset $(p_i,\ q_i)_{l=1}^{Z}$, where 'Z' is the total data size, p_i denotes input vector, and q_i denotes output vector. Equation (6.2) represents the general linear regression model.

$$f_{v,a}(p) = vp + a \tag{6.2}$$

$f_{v,a}(p)$ is a linear mixture of attributes of instance p, v is an N-directional tensor, and a is an actual numerical integer.

Classification Tree (CT)

This model has a tree form that is used for both prediction and classification problems. The CT starts at the root node and ends at the terminal nodes. It is possible to classify each terminal node based on its content. The computational time is reduced using decision trees. Multiple classification trees can be obtained from the training data and set of features. ID3, entropy, and Gini are several approaches to obtain optimal classification trees for classification problems that can give better and precise outcomes than others. In prognostication domains, the SE is utilized to find out the most critical and useful feature, which supplies the level of defects, where a lower level of defects denotes a more powerful and competent node. SE can be formulated as in Eq. (6.3).

$$\text{SE} = \frac{1}{M}\sum_{b-1}^{M}(z_u - \mu)^2 \tag{6.3}$$

M is the total occurrences, z_u is the tagged occurrences, and μ is the average of all tagged occurrences.

Support Vector Regression (SVR)

To conclude prediction problems, SVR is an altered form of support vector machine (SVM) that uses inputs instead of simply binary outputs. Input attributes are connected to a greater level by SVR, and the prognosis of nonlinear information is made possible. It generates a subgroup of the data points that are closer to the hyperplane from the input data and diminishes the margin of the classifier. The general SVR equation is denoted in Eq. (6.4).

$$S(v) = \sum_{g=1}^{Z} (\alpha_g^* - \alpha)n \cdot (v_g,\ v) \tag{6.4}$$

α_g^* and α are multipliers for finding local maxima and minima of function, and n is the kernel method.

Neural Networks that utilizes Radial Basis Function (RBFNN)

The main intention of this model is extracted from the concept of function approximation. This model differs from other models in the evaluation process done in the hidden layers. In this model, weights are calculated by applying the Euclidean distance to the radial basis function (RBF). Equation (6.5) represents Euclidean distance, and Eq. (6.6) formulates radial basis function, respectively.

$$d_q = \sqrt{\sum_{p=1}^{Z} f_p - u_{pq}^2} \tag{6.5}$$

where 'f' represents input data, and 'u' indicates hidden neuron's weight, consequently.

$$\phi = e^{\frac{-d^2}{2\sigma^2}} \tag{6.6}$$

The radius of the Gaussian curve is denoted by $\sigma > 0$, and d is the radial distance mentioned in Eq. (6.5).

The output calculation in this model is given by Eq. (6.7).

$$O(f) = \sum_{p=1}^{M} u_p \phi \tag{6.7}$$

$O(f)$ represents the model's output, 'M' represents radial basis function's count, and u_p depicts weights, respectively.

Results and Discussion

Two separate methods were taken into consideration to carry out observations, firstly discovering and getting the best and efficient prognosis model and grades for the woman players, and then to indicate the most critical attributes influencing the efficiency of the outperformed model, i.e., the ability of the athlete for the competency considered. To forecast the abilities of woman handball players, four separate fitness events, particularly a squat jump (SJ), squat jump on toes (SJT), sprint over a 10-m distance (SP10), and a *handball sport*-skill *test (HSST) were considered.* To minimize the intricacy of the data and to improve the prognosis accuracy of the ML models, all occurrences were scaled by min–max scaling. In Eq. (6.8), the min–max scaling formula is specified.

$$N_p = \frac{M_p - \min(M)}{\max(M) - \min(M)} \tag{6.8}$$

The scaled value is denoted by N_p, the data point is given by M_p, and lowest and highest values are given by min(M) and max(M) for the respective features.

The four ML models referred to in section 3(B) were independently trained in each fitness-related case using 80% of the cumulative occurrences of the remaining 23 features. After the adjustment of the attributes, end-most measurements were made by tenfold cross-validation, and 20% of the untrained instances of the dataset were used for testing. Three primary parameters were used to analyze the performance of all the models in this experiment: R^2 score, SE, and mean AE. Using mean squared error, the R-squared score is a mathematical method employed to assess the association between examined and estimated data, which is the quantification of how far a data item appears to diverge from its expected value. In Eq. (6.9), the simple R-squared score formula is given.

$$R^2 = 1 - \frac{\Sigma(e_p - \hat{e}_p)}{\Sigma(e_p - \overline{e}_p)} \tag{6.9}$$

Examined data is denoted by e_p, the expected value is given by $\hat{e}$, and the average value of all observed data is given by $\overline{e}$, consequently.

SE is the square average of the errors that have been accumulated from the residual of the values observed and the values predicted. In Eq. (6.3), the SE formula is already given. AE is the average of absolute errors that can be quantified based on the residue of the data detected and predicted. In Eq. (6.10), the AE formula is given.

$$\mathrm{AE} = \frac{1}{n}\sum_{i=1}^{n} |x_i - x| \tag{6.10}$$

where n denotes total defects, and $|x_i - x|$ represents the AE between the data observed and expected. Inefficiency evaluation of four features, namely an SJ, SJT, SP10, and HSST, four observations were conducted individually to figure out the superior variant of the considered ML models. The performance trait of woman handball players, namely SJ, was considered in the first study to forecast the abilities of players. CT had given the worst results with an R-squared score of 0.10. SLR and SVR gave very nearer outcomes of 0.707 and 0.660 for the R-squared score. Furthermore, SVR minimized the deviation. The performance trait of woman handball players, namely SJT, was considered in the second study to forecast the abilities of players. The RBFNN gave efficient results of 0.96, 0.0042 and 0.0075 for R-squared score, SE,

Table 6.2 Results of all four ML models

Skill	Metric	RBFNN	CT	SLR	SVR
SJ	SE	**0.0013**	0.033	0.0865	0.011
	AE	**0.0033**	0.1458	0.107	0.0891
	R^2 score	**0.97**	*0.10*	*0.707*	*0.66*
SJT	SE	***0.0042***	0.030	0.018	0.020
	AE	***0.0075***	0.1153	0.1131	0.1111
	R^2 score	***0.96***	0.65	0.61	0.63
SP10	SE	**0.0034**	*0.024*	0.0094	0.012
	AE	**0.0316**	*0.1297*	0.0692	0.0925
	R^2 score	**0.86**	*0.286*	0.80	0.68
HSST	SE	**0.0033**	NA	NA	NA
	AE	**0.0674**	NA	NA	NA
	R^2 score	**0.93**	NA	NA	NA

and AE. SP10 was predicted in the third analysis, and CT (0.286, 0.024 and 0.1297) received the lowest R-squared score and the best SE and AE results.

The largest prognostication was indeed achieved by the RBFNN in the last study, which was the HSST. Table 6.2 shows the overall results of 4 ML models by considering 3 performance metrics (R^2 score, SE, and AE) against 4 skills (SJ, SJT, SP10, and HSST) of female handball players. Figures 6.1, 6.2 and 6.3 represent the graphical comparison of the four ML models against performance metrics R^2 score, *SE*, and *AE* results, respectively.

Fig. 6.1 Comparison of 4 ML models against R^2 score

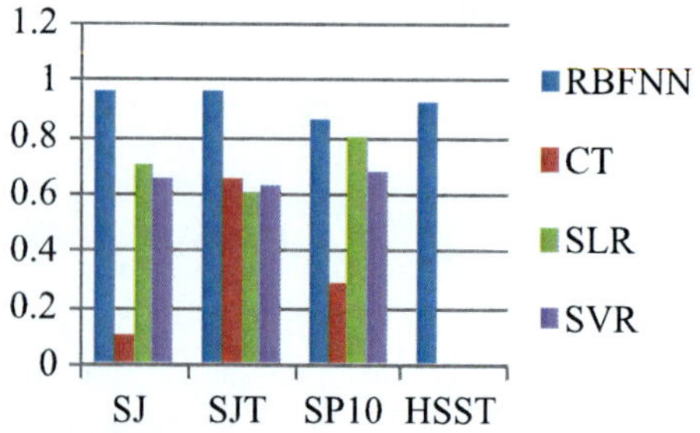

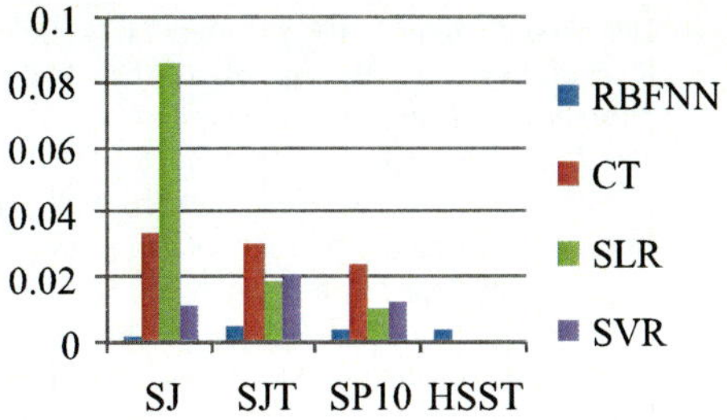

Fig. 6.2 Comparison of 4 ML models against SE values

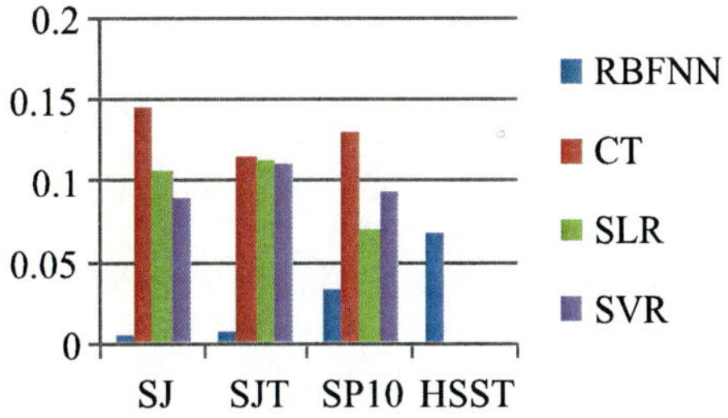

Fig. 6.3 Comparison of 4 ML models against AE values

Conclusion

Analyzing the abilities and performance of athletes before selecting a game is crucial. Athlete's performance depends on several factors and predicting which factors are important and trivial is highly impossible. Identifying the trivial parameters can help coaches to get the pros and cons of the players so that the coaches can arrange extra training sessions for weak performers. This will also help the coaches to select the best players for the final team preparation. ML models as predictive procedures are likely the most significant methods of performing expectations for complex and sophisticated tasks. Outcomes from this study demonstrated that in women handball athletes with an ML model, in particular a radial base function neural network, it is conceivable to set up nonlinear connections for different body-related and exercise limits.

References

1. Jovanovic, S., Markovic, S., Kleva, M.: Prediction of successful defense movement of female handball players. Sport Scientific and Practical Aspects. **17**(1) (2020)
2. Gomez-Lopez, M., Manzano-Sanchez, D., Merino-Barrero, J.A., Valero-Valenzuela, A.: Causes of success in handball through the beliefs about ability. RevistaInternacional de Medicina y Ciencias de la ActividadFisica y del Deporte. **20**(77) (2020)
3. Popa, D., Mindrescu, V., Iconomescu, T.M., Talaghir, L.G.: Mindfulness and self-regulation strategies predict performance of romanian handball players. Sustainability **12**(9), 3667 (2020)

4. Hermassi, S., Souhaiel, C.M., Fieseler, G., Schulze, S., Irlenbusch, L., Delank, K.S., Schwesig, R., Hoffmeyer, B.: Validity of new handball agility test: association with specific skills and muscular explosive determinants of lower limbs in young handball players. DRASSA J. Developm. Res. Sport Sci. Activities. **3**(1), 79–103 (2017)
5. Sekulic, D., Foretic, N., Gilic, B., Esco, M.R., Hammami, R., Uljevic, O., Versic, S., Spasic, M.: Importance of agility performance in professional futsal players; reliability and applicability of newly developed testing protocols. Int. J. Environ. Res. Public Health **16**(18), 3246 (2019)
6. Soslu, R., Ozkan, A., Goktepe, M.: The relationship between anaerobic performances, muscle strength, hamstring/quadriceps ratio, agility, sprint ability and vertical jump in professional basketball players. BedenEgitimiveSporBilimleriDergisi **10**(2), 164–73 (2016)
7. Csato, L.: Optimal tournament design: lessons from the men's handball Champions League. J. Sports Econ. **21**(8), 848–868 (2020)
8. Trejo Silva, A., Camacho Cardenosa, A., Camacho Cardenosa, M., Gonzalez Ramirez, A., BrazoSayavera, J.: Offensive performance under numerical inequality during exclusions in female handball. RICYDE. RevistaInternacional de Ciencias del Deporte. **62**(16), 396–409 (2020)
9. Pobar, M., Ivasic-Kos, M.: Active player detection in handball scenes based on activity measures. Sensors **20**(5), 1475 (2020)
10. Martinez-Rodriguez, A., Martinez-Olcina, M., Hernandez-Garcia, M., Rubio-Arias, J.A., Sanchez-Sanchez, J., Sanchez-Saez, J.A.: Body composition characteristics of handball players: systematic. In: Camp 2020 Jan, pp. 52.
11. Bonnet, G., Debanne, T., Laffaye, G.: Toward a better theoretical and practical understanding of field player's decision-making in handball: A systematic review. Movement Sport Sci. **4**, 1–9 (2020)
12. Groll, A., Heiner, J., Schauberger, G., Uhrmeister, J.: Prediction of the 2019 IHF World men's handball championship–a sparse Gaussian approximation model. J. Sports Anal. **6**(3), 187–197 (2020)
13. Niksic, E., Beganovic, E., Joksimovic, M.: The impact of the program of basketball, volleyball and handball on the situation-motorized capability of the first classes of the elementary school. Pedagogy of Physical Culture and Sports **24**(2) (2020)
14. Trebinjac, S., Nair, M.K.: Injury mechanisms in sports. In: Regenerative Injections in Sports Medicine, pp. 7–16. Springer, Singapore (2020)
15. Pescatello, L.S., Riebe, D., Thompson, P.D., (eds).: In: ACSM's Guidelines For Exercise Testing and Prescription. Lippincott Williams & Wilkins (2014)
16. Bar-Or, O.: The Wingate anaerobic test an update on methodology, reliability and validity. Sports Med. **4**(6), 381–394 (1987)
17. Iacono, A.D., Eliakim, A., Meckel, Y.: Improving fitness of elite handball players: small-sided games vs. high-intensity intermittent training. J. Strength and Conditioning Res. **29**(3), 835–843 (2015)
18. Moir, G.L.: Three different methods of calculating vertical jump height from force platform data in men and women. Measurem. Phys. Educ. Exercise Sci. **12**(4), 207–218 (2008)

Chapter 7
Deep Learning for Object Detection: A Survey

Mohan Krishna Kotha and Kanadam Karteeka Pavan

Introduction

To find all the locations of items of interest in an input, object detection uses a task that uses bounding boxes and labels them into categories to which they belong. Several ways have been proposed to accomplish this goal, ranging from classical methods to those based on deep learning. Techniques to object detection may generally be divided into two categories: those based on region proposal algorithms, which are referred to as two-stage approaches and those relying on regression or classification, which are known as real-time and unified networks, or one-stage approaches. Real-time object detection applications have gained a lot of traction recently due to the need to satisfy the demands of modern living while also helping people live better. Self-driving vehicles, for example, are a real-world illustration of how they might assist people travel securely on roadways while lowering the number of accidents caused by distracted drivers. Detecting faulty assembly components or angle of view, deformable shape and size of detected object changes that are significant during the assembly process are important in manufacturing businesses. Thus, it demonstrates the importance of real-time object recognition in popular vision-based practical applications. However, early object identification is required for these applications to be employed as inputs in later activities. Early detection means that representations of items are small or even invisible due to their small size.

Object detection has recently gotten a lot of interest because of the variety of applications it may be used for and the recent technological advances that have made it

M. K. Kotha (✉)
Research Scholar, Acharya Nagarjuna University, Guntur, Andhra Pradesh, India
e-mail: mohankrishnakotha@gmail.com

K. K. Pavan
R. V. R & J. C College of Engineering, Chowdavaram, Andhra Pradesh, India

Ch. Satyanarayana et al. (eds.), *Proceedings of the International Conference on Computer Vision, High Performance Computing, Smart Devices and Networks*, Advanced Technologies and Societal Change,
https://doi.org/10.1007/978-981-19-4044-6_8

possible. Drone scene analysis, autonomous driving, transportation surveillance and robotic vision are just a few of the areas where researchers are looking at this problem in depth in both academia and the real-world. The rapid advancement of object identification algorithms may be credited to a variety of variables and initiatives, but the development of deep convolution neural networks and the computational power of GPUs stand out. Computer vision in general, as well as object recognition specialized to a certain domain, has embraced a deep learning methodology in recent years. Deep learning networks are commonly used as the backbone and detection network in state-of-the-art object detection systems to extract features from input images, classify objects and locate them. Visual recognition is a computer technique that identifies instances of meaningful items in digital photos and videos. It is linked to computer vision and image processing. Multi-category detection, edge detection, salient object detection, face detection, scene text detection, posture detection and pedestrian detection are among the well-researched object detection areas.

Backbone Networks

Backbone is a cornerstone functionality for target identification. The last completely associated layers of the neural network are taken in the method of object recognition. There has been an upgraded version of the classification network now open for use. The same as the first author, Lin et al. [10] complement or delete layers or cover all of the layers with special build content. Any works use the fresh bones for extraction function. We can develop and differentiate deeper and tightly connected neural networks, such as [8, 11–17]. Lightweight backbones will satisfy expectations for smartphones. Wang et al. [18] proposed a novel object tracking method by integrating [7] with SSD [2] and optimizing the architecture for high-speed dispensation. In instruction towards satisfying high precision and more reliable demands, dynamic media backbone has to be used. Many computers have this feature such that they can be used to retrieve digital information over the network at high speed without compromising on the quality of the information. To prove the more accurate detecting of land distribution, a denser and more connected backbone is preferred over a shallower also sparse particular one. He et al. use ResNet rather than VGG for maximizing recognition performance because of its higher precision. More accurate and precise classifiers can be useful for exact object detection tasks. This will contribute to improving the efficiency of the whole Internet since the backbone system serves as a function extractor. It is noticed that the excellence of edges decides the higher limit of system efficiency. Hence, it requires further examination.

Two Kinds of Object Detectors

For the most part, pre-existing domain-specific image object detectors fall into one of two categories: the two-stage detector or Faster R-CNN [5]. YOLO [6] and SSD [7] are examples of one-stage detectors. One-stage detectors achieve high inference speed while two-stage detectors have elevated localization and object recognition accuracy. The ROI Pooling layer divides the two stages of two-stage detectors.

Characteristic Baselines

In recent years, there has been a lot of advancement in the arena of object recognition. The invention of R-CNN, a CNN-based object indicator, has resulted in major progress in the arena of object detection. This segment will show you a few examples of object detection architectures.

2-Stage Detectors

(1) CNN detector based on zone dependent. Girshick's et al. [20] have created an approach called R-CNN which improves object detection efficiency by a very considerable amount. Deep learning is very quick and effective. The R-CNN has four modules. The first module functions independently without submission to colleagues. Module two extracts vector features in each proposal. These are linear SVMs for the shapes in a single image. The last module will help in predicting the correct bounding boxes with reliability. Writers create ideas in metropolises. Then CNN feature is derived from each field proposal. The floorplan of the region features should have the same size. The writers determined 227 pixels in size for the input text in CNN. Each picture has a unique aspect. Regardless of the size of the area to be analyzed, writers warp all pixels around the candidate area. We have five convolutional and two completely connected layers. All CNN variables are distributed over all local variables. We train each separate model separately.

(2) To fine-tune deep convolutional neural networks, we should make use of a broader dataset. Girshick et al. (2012) train a large-scale CNN arranged a historical dataset (ImageNet classification dataset [3]). The last layer remains upgraded to become a standard CNN that can be used for specific tasks like image classification. To deal with the discrepancies of CNN, SGD is used to update the coefficients from the proposed structure windows. There is an (N + 1)-way grouping layer above it (N: objects classes, 1: background). The writers can discuss situations, whether positive or negative, in two ways: positive scenarios and negative scenarios. In the fine-tuning step, it is necessary to specify the connection ended union (IoU) overlap verge by way of 0.5. Currently, only zone proposals that are below the threshold and object proposals are included in the election process. Contrary to plans for or plans which (Fig. 7.1).

An alignment through a ground-truth class is allocated towards each case. Another circumstance is to categorize the data when assigning SVM. In comparison, ground-truth class voting is performed considering only positive ground-truth examples of one session. These ideas match from 0.50 and 1 which increases the number of instances from effective ones by 30%. It effectively avoids overfitting during fine-tuning process. Faster edition of R-CNN has proposed sometime after. R-CNN produces extremely pricy plans without exchanging the computer tools and resources

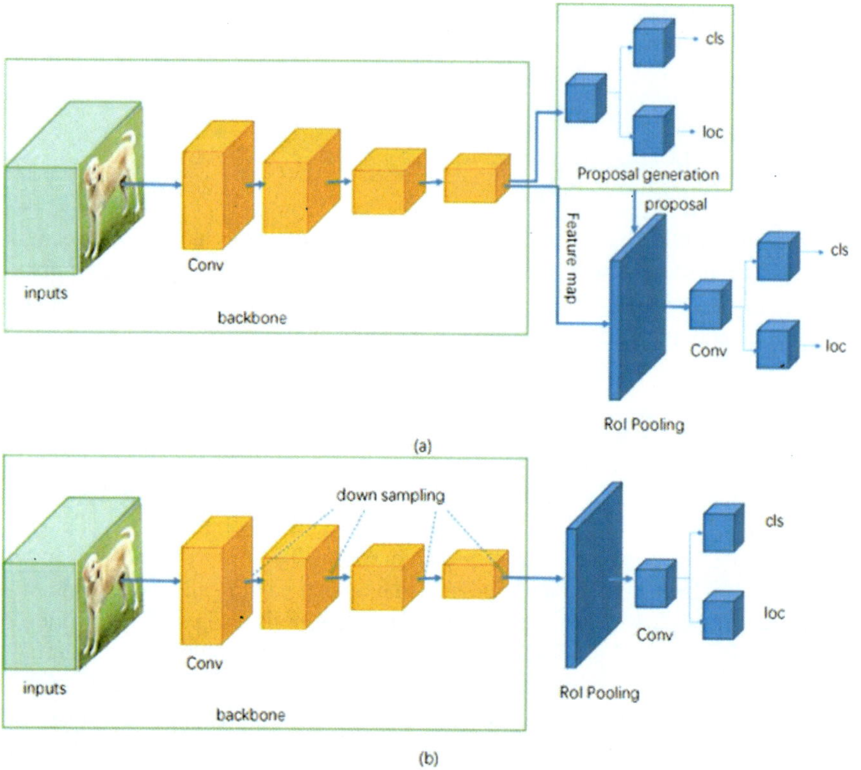

Fig. 7.1 **a** To feed region suggestions interested in the classifier also regressor, an area proposal network is used. **b** illustrates how the computer predicts a bound in g box from the image by solving a linear equation. Technically, yellow cubes are one block with the same resolution and implementation of convolutional layers on each sheet. A sequence of convolutional layers is featured by thick blue cubes. Flat Blue CUBE shows the RoI Pooling sheet

with other proposed ideas. It remains recommended to combine a feed-forward and recurrent neural network towards function as a classification network. These features are automatically removed from the image until they are reported to CNN in order to do the classification process and to do the localization of objects. Contrasting. R-CNN can be used to quickly and efficiently handle large datasets. Training period of ReLuC consists of pre-training, fine-tuning and final bounding box regression. The first stage of R-CNN runs on a sequence of single image segmentation followed by object detection. The fixed size function maps of Fast R-CNN are dynamic. This approach saves hard disc space. For more accurate detection, the author uses the truncated SVD. It was discovered that Fast R-CNN has about the similar accuracy percentage by way of R-CNN. Cost of preparation of R-CNN has fallen to 70 hours on previous ANN The R-CNN model was 213 times faster than DeepLab. (47s). This has been done through experimentation on Nvidia K40 GPU.

Fast R-CNN remained then modified into Faster R-CNN and the feature extraction process was better quality. Fast R-CNN remains as same as detector network with less modifications. Faster R-CNN uses a fast are a suggestion network that works well on videos of different sizes and aspect ratios. RPN accelerates output image generation as it can be combined into the entire detection network. The information in this figure explains the protocol in Fig. 7.3b. Alternative way to map moving objects is to use multi-scale reference data. Anchors will greatly simplify the task of generating plans for different kinds of the regions due to the availability of maps containing features of independent and dependent variables. Then, moving a 3×3 window over the activation map, the centre of the window is changed relative to the initial input's centre point. The authors explain three different sizes of anchor boxes. The location (the area) is defined relative to the anchor box. They also measured the distance from the predicted box to the correct value box.

Experiments have shown that Faster R-CNN has dramatically increased accuracy and tempo. Faster R-CNN obtained a score of 69.9% associated towards Fast R-CNN which obtained 66.9% on same 2007 Mathematical Assessment and Curriculum Authority Much Faster R-CNN runs even slower than Fast R-CNN, through the similar backbone. Mask R-CNN. For instance sectioning for Faster R-CNN. Regardless, Mask R-CNN is able towards identify targets. He et al. used ResNet with Fast R-CNN. FPN (cylinder scale). This algorithm will achieve high precision and high speed. The act combines top-down control as well as bottom-up control. Component Networks can generate feature hierarchy from two lower levels. Hierarchical upsampling at the pyramid level results in superior photographs. The last convolutional layer will snatch the last top-level attribute charts. There will be convergence of the internal states of both cases. Both shape and location of feature maps remain distinct, the bottom of the convolutional coat can change the position of the feature map. Since the connection was formed, the top floor had transformed into a pyramid instead of a square. It's essential to have features such as good contrast in this case.

A. Another feasible way to score higher is to merge dense feature maps separately. Traditional pooling does discrete numbers pooling in two stages to determine each function value. First, quantization is used to evaluate the coordinates of features in the pictures. Then, downsampling is needed. Then function maps of ROI are separated to the binary bins, wherein the size is quantized. From the two quantization operations of several raw images, misalignment was the result. In this way, RoAlign stops and prevents the accurate measurement of Return On Investments borders or bins. First, the software utilizes quadtree techniques to determine the precise positions of the features in each rural area's diagram. It then combines the values of each bin to form an aggregated value for the whole bin. The ROIAlign mechanism as seen in Fig. 7.2.
B. Experiments have shown that with the aforementioned two changes, accuracy improved. Using ResNet-FPN, one can get up to 75% accuracy arranged MS COCO dataset.

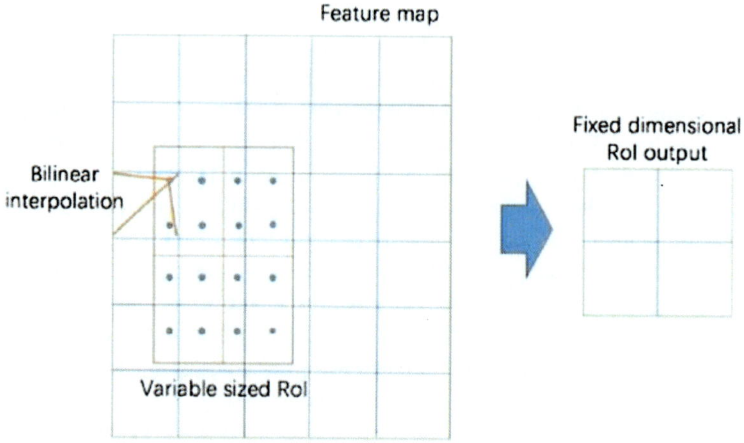

Fig. 7.2 Aligned operations. The first move determines object's position in the function diagram. Finally, the values of the features would be derived from sampling at four sites, in the divided bin

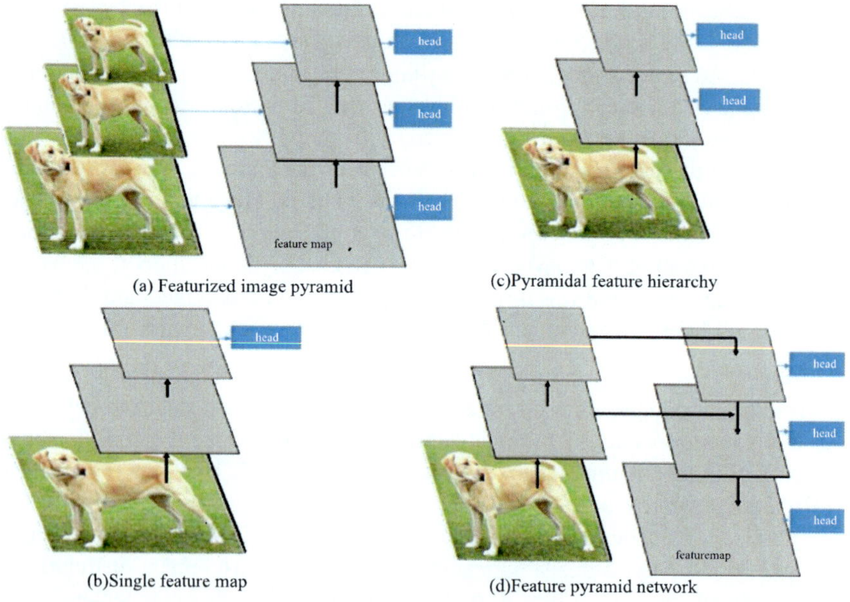

Fig. 7.3 There are four approaches to handling prediction of various-sized items. Use images to create a pyramid of facts. Features are computed on different scales, which is sluggish. **a** and **b** ConvolutionalNet (23) returns a pyramid of features, **b** is lightfast, however **c** is more precise

C. **One-stage Indicators**

YOLO [6] remains a CNN-based target sensor suggested through Redmon et al. after FRC [5]. The key influence remains eliminating the effects of partial descriptions on detection tasks. Firstly, r-CNN overfits fewer bounding boxes in the picture compared to Fast R-CNN.

Secondly, detection frame from the dilemma is taken from the results of regression. On a Titan X GPU, P (object) indicates how likely it is that the box you specified contains an object and IOU (intersection over union) indicates how real it is that the box you specified contains an object. The B bounding boxes and C class predictions of corresponding categories are predicted by each rectangular cell in the grid. There are 24 convolutional layers in the network, followed by two completely connected layers. The first 20 convolutional layers and a pooling layer are used for pre-training on the ImageNet database by the authors. Both analogue and digital systems had its own merits. To get fine-grained information for better identification in the detection phase, half the resolution, which is 224 224 would be used in pre-training.

(1) YOLOv2 [22] is a follow-up to YOLO [6], which incorporates a collection of novel concepts to improve YOLO's pace and knowledge. Serialize your data. The downstream layers can benefit from a fixed distribution of inputs to a convolutional layer. Since the second stage uses a stochastic gradient descent optimization technique, it cannot be started. SGD employs mini-batches, which cuts down on preparation time. Batch runs a number of statistical tests on the information. After computing the data's mean and variance, the activations of number m are normalized to have a mean of zero and a standard deviation of one. The mini-batch components come from the same distribution as the main batch. This can be thought of as a BN layer that produces activations at the same frequency. In addition, placing a BN layer ahead of each convolutional layer speeds up the convergence process and helps to regularize the model. In mAP, batch normalization reduces mAP by more than 2%. A classifier is a device that categorizes objects The YOLO classifier uses a 224 × 224 × 448 input resolution until increasing it to 448 × 448 for detection. When you move to an object detection mission, you'll need to reconfigure this network for a new resolution.

Convolutional Encoder Networks are a form of convolutional encoder. Initial YOLO networks generate the predicted box's coordinates directly. Faster R-CNN produces detected boxes by using anchor boxes as temporary detectors. YOLOv2 starts with removing layers that include completely linked components. And it creates predictor the classes and objects for each anchor package. This action improves recall by 7% and also reduces total mAP by 0.3%. Using information from primary dimensions to predict the length and width of box, for calculation of the aspect ratio. High-Performance The use of high-resolution features is useful for localized small artefacts. YOLOv2 concatenates low-resolution features with higher-resolution features, similar to ResNet's database, by repeating features into different channels for a limited performance gain. Multi-level education. To maintain parallel effectiveness, the network selects a new image dimension size from 320, 352 and 608 every 10 cycles. This implies that the same network can be used to model fluid motions at various scales. On the VOC 2007, YOLOv2 achieves 78.6% mAP and 40fps, compared to 63.4% mAP and 45fps for YOLO. The YOLOv2 model is optimized for Darknet-19 classification, which requires less operations and achieves high precision. Faster R-CNN with ResNet backbone and SSD500 has a lower mAP and fps than the more competitive YOLOv2 version. With modern complex architecture, YOLOv2 achieves high detection accuracy despite some disadvantages.

(1) YOLOv3 uses three separate function maps in predicting the bounding box. The final convolutional layer produces 3D tensor decoded projections and bounding boxes.
(2) It was found that YOLOv3 (AP:33.2%) delivers an equivalent output to SSD (AP:33%) though three times faster on the same dataset (AP:40.8%). Despite using the old-school metric of mAP at IOU $= 0.5$, YOLOv3 could attain 57.9% mAP as opposed to DSSD513 of 53.3% also RetinaNet of 61.1%. Thus, YOLOv3 has comparatively inferior prediction ability to prioritize small items.

SSD forecasts individual group scores and box-offsetting scores in a single shot with a given range with avoidance. (a). The bounding boxes in each element have various aspect ratios and sizes. If done well, this method is flexible and open to various types of artefacts, features, etc. Figure 7.3c indicates the proposed solution. At training time, the system has to pick out the ground reality boxes matched to the default boxes to serve as positive indicators of the processing. The writers will select which three negations are the most convincing and then fill up the box accordingly. The authors do data augmentation which is a very efficient method to make large improvements in the precision of EDA.

In addition, experiments on the PASCAL VOC 2007 and MS COCO datasets showed significant results in the model growth.

The authors suggest focusing loss function to discount losses on difficult cases. These difficult cases are much easier to spot and can help minimize false positives, completely overcomes the downside of the earlier one-stage detectors. In comparison to MS COCO test-dev datasets, ResNet-101-FPN backbone received 39.1% AP. ResNet-101-FPN is a ResNeXt-101-FPN 40.8% AESCAP well above that of DSSD513. Zhao et al. [26] build a multi-level function pyramid network to generate a variety of features across object instances (MLFPN). To arrive at the final improved feature structure, the developers go through three stages. The base element is formed by fusing the features extracted from several layers in the backbone. Second, the base part is fed into a block made up of joint thinned U-shape modules and feature fusion modules in alternating order. Basically, a function pyramid is built by combining the decoding layers with equal scales. Currently, multi-scale and multi-level is being planned. Part of the research is to adopt the SSD (Software system design process), to receive.

M2Det remains a one-stage detector and hence it accomplishes the output of 41.0 at rate of 11.8 FPS through one-step approach. It is slower than RetinaNet800 through 0.9% with single-scale inference strategy, but it outperforms RetinaNet800 by 0.9%.

(2) RefineDet: RefineDet [28] involves two integral subsystems: the anchors' refinement subsystem and the entity recognition subsystem. The new model is linked to the old model since it shares characteristics with the older model. Training is performed through three phases, pre-processing, detecting and prediction. Classical on-stage detectors like SSD, YOLO, RetinaNet both use on-stage classification to achieve final data. Two-phase cascaded regression method is more efficient in identifying hard-noticed substances, particularly for unimportant objects.

(3) ***Relation Networks for Object Detection***: Hu et al. proposed an adaptation of focus mechanism called interaction matrix for object detection [29]. The object sensing module is located before two completely functioning layers to increase the accuracy of object identification. Besides directing all AI features into a supervised classification or regression model, the link module often takes care of the anomaly detection mechanism itself. Through by means of Faster R-CNN, FPN then DCN as the backbone system happening the COCO files, the connection detection results will be improved by 0.2%, 0.6% and 0.2%, respectively.

(4) ***DCNv2***: The writer recommends DCNs for handling the significant variation in spatial information in the visual stimulus. Regular ConvNets can only recognize and classify visual features of fixed pixel sizes (e.g. according to the kernel).

(5) In comparison to DCN v1, Adjustment was induced as a result of training. Function Spot Matching was used to boost the process efficiency. DCNv2 achieves 43.0 % mAP, while DCNv1 achieves 41.8 % mAP. The mAP signifies Mean Average Precision. On the other hand, DCNv2 improved R-identification CNN's score by 4.6% – 5.2% and 6.5% – 9.1% faster than FCNs, respectively.

(6) Google Brain continued to investigate the field of neural architectures in pursuit of new possible architectures that could also search for new functionality among other architectures. During the recurrent rounds, the high-level mechanisms mimic the low-level modules until the architecture is complete. Therefore, for maximizing the level of the accuracy, high resolve feature maps should be created aimed at minor detectors. Both the above characteristics would certainly improve the precision of identification. On COCO test-dev dataset, ResNet-50 performed the best, which performed better than the original FPN. The superlative configuration of NARS-FPN obtained 44.2% validation COCO test development. In conclusion, the standard baseline extraction enhances the precision of object detection by removing further features from low-resolution images and creating multi-level also multi-scale features in higher-resolution images. To speed up the search process, the one-stage detectors screen out simple samples, which exclude a significant portion of proposals. To adapt to geometric variation, it is possible to use deformable layers. This can play a big role in improving

computer efficiency. Results on exam settings of MS COCO tests of standard starting point are described in Table 7.2.

Dataset Salsometrics

To classify anybody, one must be able to detect its location. Bounding boxes, as seen in Fig. 7.4, may be used to depict an object's location. Early algorithms were based on face recognition using simple techniques. Yet more realistic datasets of facial expression were then developed. A current and familiar difficulty is on the identification of pedestrians. The Caltech Pedestrian Dataset (CPD) has 350,000 instances with location information. These measurements are useful to evaluate the efficiency of detectors through conforming outputs.

PASCALVOC Dataset
Datasets used in the experiment are commonly used in the country. The 2008 Pa VOC datasets (VOC2007) contain over 11,000 pictures for each category. There are 20 major categories for the classifications; they include cars, livestock, household items and persons. Approximately of them change the semantic specificity of the production, such as car also bike, auto, motorcycle and motorbike, but not look alike. Furthermore, facial similarities between the groups raise the complexity of identification, e.g. dog versus cat. Out of the 27,000 object instances springing frames, about 4000 have extensive segmentations.

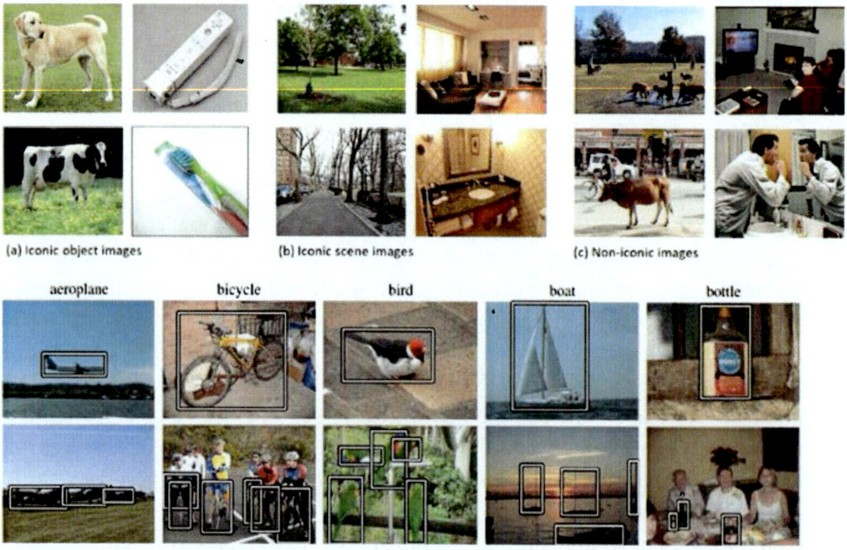

Fig. 7.4 There are two lines after the MS COCO dataset. The photographs depict three changed categories of images sampled in the dataset

MSCOCO Benchmark

(1) **Dataset**: Microsoft Generic Objects in Context (MS COCO) to detect and segment objects encountered in daily life in their natural environments. These 20 labels cover the 20 Pa groups. In total, the dataset includes 2.5 million labelled examples and 328,000 photographs. The main feature of MS COCO dataset is that it provides full survey of diverse artefacts in context. In comparison to the ImageNet database (Table 7.1).

Comparing with output of VOC and ImageNet. (3.0). More importantly, MS COCO has several categories per image as opposed to ImageNet on average. Besides 10 percent images fashionable, MS COCO consumes just one category, PASCAL VOC and ImageNet have additional than 60% images of only a single group. Unrecognized items can be more difficult to remember. Images provide rich contextual details. More than 800,000 class is the guy while 600 class is hair carrier. Another such limited number is class of hair brushes with a number above 800 Besides.

(2) ***Metric***: The MS Cocos metric is very strict and evaluates the success of identification fully Aside from that, basic test specimen properties are segmented in order to measure the accuracy of the. This optical lens was able to differentiate between objects of various sizes.

ImageNet Benchmark

(1) ***Dataset***: The test datasets will spur a step into developing real-life use cases. Another crucial large-scale benchmark dataset is the ImageNet database [3]. The ILSVRC challenge tests the ability of an algorithm to locate objects in an image. ILSVRC 2014 has more than 200 object types and almost 450,000 teaching photographs, 20,000 validation images and 40,000 evaluation images. There are similarities between UbiComp and PASCAL VOC.
(2) ***Metric***: The threshold $t = 0.5$ is used in the PASCAL VOC metric. Small object deviations of a few pixels.
(3) ***VisDrone2018benchmark***: VisDrone2018 [4], a large-scale dataset of photographs and videos collected by drones, was released last year.

Digital object monitoring benchmark dataset. This dataset advances visual awareness activities with the aid of drone. The photographs and video sequences were taken from numerous urban and residential areas from north to south. This is a benchmark dataset that facilitates deeper study of optical tracking and recognition algorithms on drone platforms. Due to a vast number of small items such as dense vehicles, pedestrians and cyclists, it would be impossible to detect specific groups the visibility of the photographs is tiny, which causes the difficulty of detecting small and dense objects. Figure 7.5 shows this. The dataset I use adopts Microsoft MS COCO metric.

Table 7.1 According to the inferencing results of certain typical baselines

Method	Data	Backbone	AP	AP_{50}	AP_{75}	AP_S	AP_M	AP_L
FastR-CNN [29]	train	VGG-16	19.7	35.9	–	–	–	–
FasterR-CNN [8]	trainval	VGG-16	21.9	42.7	–	–	–	–
OHEM [40]	trainval	VGG-16	22.6	42.5	22.2	5.0	23.7	37.9
ION [41]	train	VGG-16	23.6	43.2	23.6	6.4	24.1	38.3
OHEM++ [40]	trainval	VGG-16	25.5	45.9	26.1	7.4	27.7	40.3
R-FCN [42]	trainval	ResNet-101	29.9	51.9	–	10.8	32.8	45.0
CoupleNet [43]	trainval	ResNet-101	34.4	54.8	37.2	13.4	38.1	52.0
FasterR-CNNG-RMI [44]	–	Inception-ResNet-v2	34.7	55.5	36.7	13.5	38.1	52.0
FasterR-CNN+++ [25]	trainval	ResNet-101-C4	34.9	55.7	37.4	15.6	38.7	50.9
FasterR-CNNwFPN [15]	trainval35k	ResNet-101-FPN	36.2	59.1	39.0	18.2	39.0	48.2
FasterR-CNNwTDM [45]	trainval	Inception-ResNet-v2-TDM	36.8	57.7	39.2	16.2	39.8	52.1
DeformableR-FCN [38]	trainval	Aligned-Inception-ResNet	37.5	58.0	40.8	19.4	40.1	52.5
*umd*_det [46]	trainval	ResNet-101	40.8	62.4	44.9	23.0	43.4	53.2
CascadeR-CNN [47]	trainval35k	ResNet-101-FPN	42.8	62.1	46.3	23.7	45.5	55.2
SNIP [48]	trainval35k	DPN-98	45.7	67.3	51.1	29.3	48.8	57.1
Fitness-NMS [49]	trainval35k	ResNet-101	41.8	60.9	44.9	21.5	45.0	57.5
MaskR-CNN [11]	trainval35k	ResNeXt-101	39.8	62.3	43.4	22.1	43.2	51.2
DCNv2 + FasterR-CNN [39]	train118k*	ResNet-101	44.8	66.3	48.8	24.4	48.1	59.6
G-RMI [44]	trainval32k	EnsembleofFiveModels	41.6	61.9	45.4	23.9	43.5	54.9
YOLOv2 [30]	trainval35k	DarkNet-53	33.0	57.9	34.4	18.3	35.4	41.9
YOLOv3 [32]	trainval35k	DarkNet-19	21.6	44.0	19.2	5.0	22.4	35.5

(continued)

Table 7.1 (continued)

Method	Data	Backbone	AP	AP_{50}	AP_{75}	AP_S	AP_M	AP_L
*SSD*300* [10]	trainval35k	VGG-16	25.1	43.1	25.8	6.6	22.4	35.5
RON384+++ [50]	trainval	VGG-16	27.4	49.5	27.1	–	–	–
SSD321 [34]	trainval35k	ResNet-101	28.0	45.4	29.3	6.2	28.3	49.3
DSSD321 [34]	trainval35k	ResNet-101	28.0	46.1	29.2	7.4	28.1	47.6
SSD512* [10]	trainval35k	VGG-16	28.8	48.5	30.3	10.9	31.8	43.5
SSD513 [34]	trainval35k	ResNet-101	31.2	50.4	33.3	10.2	34.5	49.8
DSSD513 [34]	trainval35k	ResNet-101	33.2	53.3	35.2	13.0	35.4	51.1
RetinaNet500 [33]	trainval35k	ResNet-101	34.4	53.1	36.8	14.7	38.5	49.1
RetinaNet800 [33]	trainval35k	ResNet-101-FPN	39.1	59.1	42.3	21.8	42.7	50.2
M2Det512 [35]	trainval35k	VGG-16	37.6	56.6	40.5	18.4	43.4	51.2
M2Det512 [35]	trainval35k	ResNet-101	38.8	59.4	41.7	20.5	43.9	53.4
M2Det800 [35]	trainval35k	VGG-16	41.0	59.7	45.0	22.1	46.5	53.8
RefineDet320 [36]	trainval35k	VGG-16	29.4	49.2	31.3	10.0	32.0	44.4
RefineDet512 [36]	trainval35k	VGG-16	33.0	54.5	35.5	16.3	36.3	44.3
RefineDet320 [36]	trainval35k	ResNet-101	32.0	51.4	34.2	10.5	34.7	50.4
RefineDet512 [36]	trainval35k	ResNet-101	36.4	57.5	39.5	16.6	39.9	51.4
RefineDet320+ [36]	trainval35k	VGG-16	35.2	56.1	37.7	19.5	37.2	47.0
RefineDet512+ [36]	trainval35k	VGG-16	37.6	58.7	40.8	22.7	40.3	48.3
RefineDet320+ [36]	trainval35k	ResNet-101	38.6	59.9	41.7	21.1	41.7	52.3
RefineDet512+ [36]	trainval35k	ResNet-101	41.8	62.9	45.7	25.6	45.1	54.1
CornerNet512 [51]	trainval35k	Hourglass	40.5	57.8	45.3	20.8	44.8	56.7
NAS-FPN [18]	trainval35k	RetinaNet	45.4	–	–	–	–	–

(continued)

Table 7.1 (continued)

Method	Data	Backbone	AP	AP_{50}	AP_{75}	AP_S	AP_M	AP_L
NAS-FPN [18]	trainval35k	AmoebaNet	48.0	–	–	–	–	–

EPI, EPI, EPI 25 Scores
* indicates significant dataset

Table 7.2 Comparing amongst ILSVRC object detection dataset also Pascal VOC dataset

Dataset	Classes	Fully annotated training images	Training objects	Val images	Val Objects	Annotated objects
PASCAL VOC	20	5717	13,609	5823	15,787	2.7
ILSVRC	200	60,658	478,807	20,121	55,501	2.8

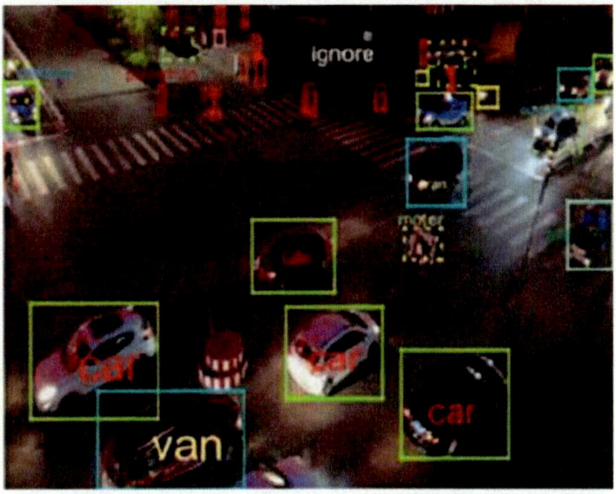

Fig. 7.5 An object image without bounding box or group name label. The pictures are of the drone from Visdrone 2018 dataset [4]

OpenImagesV5

(1) ***Dataset***: Open images version 5 is made of 16 M since all of the boxes have been made by professionals. This television series is full of varied and complicated sequences (8.3 per image on average). There are annotations in this data collection that show things that are logically related to each other. In all, there are 329 triplets and 391,073 tests. Fourth, the V5 segmentation process supplies segmentation. Segmentation masks define a much greater volume of both the objects and their spatial relationships to one another. Finally, the dataset contains 36.5 M image-level labels covering 21,973 groups.

(2) ***Metric***: According to fair outcome, the un-annotated classes are left out to avoid wrongly countermanded votes. Secondly, if an object belongs to a certain class, then For each class the subclass models should give positive prediction. If there was a missed case in one of the classes, it would be known as a false negative. There also exist groups of a single object instance occluded or touching each other in Open Images Dataset. If an individual is detected inside the package and the detection area is greater than 0.5, this would constitute a positive detection. True positives only count if they are detected in the same group of face.

Examination of General Image Object Recognition Approaches

The framework of Deep Neural Network (DNN) based object recognition involves four stages, pre-processing, feature extraction, classification also localization and post-processing. Moreover, these images are not usable in the training. We need to scale the photos to particular size to make the picture as plain and precise as possible. It is available to meet your requirements and handling of resolutions. GANs (generative adversarial networks) can generate new images to increase the sampling diversity. Please see [50] for details about data augmentation. Detection is important in function extraction. The accuracy of a dataset explicitly affects the ability to be able to do the next task including classification and localization. The responsibility for proposing the bounding box boundaries and final scores is shouldered by the detector head. There are several approaches to be used individually or in conjunction with each other to get accurate detection results.

A. Enhanced features

The salient feature extractor is the real backbone for further analysis (Tables 7.3 and 7.4).

To extract various amounts from successive backbone layers, Lin et al. [10] attempt to work out how to cut them into different levels (d). Any models are created using the Open ACC power pyramid. This is done in this fashion with the assistance of eye sensor. Peng et al. [52] consider expanding the feature-wise connectivity networks instead of the depth-wise ones. This feature extraction technique is usually part of a computer Programme. Li et al. (2012) will combine features to achieve final answers. Chen et al. [55] suggested WeaveNet, an adaptive system to weave context knowledge across scales and allow more accurate context reasoning. Zheng et al. need to retain background knowledge for the shallow layers of the visual encoder. Semantic relationships between the target borders in an image can be helpful in detecting certain small objects. Bae et al. use many sophisticated features to classify and localize. Zhang et al. [28] emphasized semantic features of object clusters into the contents of the deep detector. Scene context relationships can be helpful in face recognition. LUCC defined that Liu et al. [59] find the qualitative explanations to boost their result. If we model objects then we can get object detection capabilities. It can be done with the sub-region selection algorithm. According to Hu and colleagues, there is a linear structure of objects that defines objects' presence concurrently. This process can be done with geometric features of interaction. Any visual properties of objects can impair object detection [34].

The attention mechanisms will function well with connections looking at the least important area. Some studies [73–79] take into account the approach process to discover more simple approaches to detect an item. A combination of global interest and local reconfigurations can collect a variety of information. The use of the strong area of one object will lend accuracy. Original networks only concentrate on local characteristics, including width of the cube, which ignores the entire

Table 7.3 Via the (intro-) text, I had the impression that person detection benchmarks

Dataset	Countries	Cities	Seasons	Images	Pedestrians	Resolution	Weather	Train cal test split (%)
Caltech [1]	1	1	1	249,884	289,395	640 × 480	Dry	50-0-50
KITTI [2]	1	1	1	14,999	9400	1240 × 376	Dry	50-0-50
CityPersons [44]	3	27	3	5000	31,514	2048 × 1024	Dry	60-10-30
TDC [45]	1	1	1	14,674	8919	2048 × 1024	Dry	71-8-21
EuroCity Persons [43]	12	31	4	40,217/7118*	183,004/35,309*	1920 × 1024	Dry/wet	60-10-30

* indicates significant dataset

Table 7.4 Assessment of pedestrian detection datasets

Dataset	Imaging setup	Pedestrians	Neg images	Pos images	Pedestrians	Neg images	Pos images
Caltech [1]	mobile	192 k	61 k	67 k	155 k	56 k	65 k
INRIA [46]	Photo	1208	1218	614	566	453	288
ETH [47]	Mobile	2388	–	499	12 k	–	1804
TUD-Brussels [48]	Mobile	1776	218	1092	1498	–	508
Daimler-DB [49]	Mobile	192 k	61 k	67 k	155 k	56 k	65 k

The 3rd to 5th is training sets. The 6th, 7th and 8th are tests. It is important to note the information from Piotr et al. IE Plenary 2012 [2]

appearance of the object. This will generate a deformable kernel which will generate an offset. Deformable ROI Pooling can manage partial position entities in an easier way. The network weights and the sampling together decide the behaviour of the successful support field. Besides, rich, provocative imagery will enhance mission results. Biologically inspired systems and mechanisms have enabled detection to be more effective.

B. Increasing localization accuracy

Object identification includes localization, recognition and classification. Localizing objects can increase the accuracy of object detection models and can greatly improve detection efficiency. One way to improve the localization precision is to use a novel loss function. In this way, the output can be used to determine the usability of the output in detecting the object. Its ground reality box and its matching bounding box that connects with it. The calculation may be based on this formula. This would result in improved NMS monitoring procedures. One sector object detector can be paired with a sector detector. According to recent logic, network forecasting requires an IoU loss for precision in making projections. The loss function can detect objects and people in all shapes and ages in minutes. It integrates a new metric into the original object detection model to improve accuracy. Tychsen et al. [40] adopted a novel regression analysis for bounding box data. There would be unexpected variation from forecast and delegated expected value at any stage but not at the highest level. Researchers also proposed bounding box regression as learning technique for bounded box localization at the same time. Hsu and Aliprantis developed a novel bounding box regression failure method, which has a major influence on localization accuracy. Pang et al. [52] recently provided statistical balance with an additional penalty term for unintentional occlusion. Cabriel et al., in their report, proposed a means to get high localization precision at the cellular level. Researchers also developed a new location-based global climate policy division, which is supposed to make the forecasts more precise.

Applications also Branches

Key request fields:

Object detection is successfully used in the military, security, transportation, medical and life fields. We will discuss these fields in detail for you to know about them.

(1) **Security field:** Protection will be one of the biggest beneficiaries of facial and fingerprint recognition. Face detection attempts to detect a face in a picture. Recognition and imaging is also a little problematic for facial recognition systems. Many writers are concerned with the construction and modelling of precision detectors. Pedestrian detection is a technique for detecting pedestrians in natural settings. Braun et al. [44] published the Euro City Pedestrians, Cyclists and Other Riders dataset, which depicts pedestrians, cyclists and other riders in urban traffic scenes. Pedestrian of analytic cascade. Automatic object detection systems. Another group of researchers is concerned with automated and precise target findings. Shahzad et al. in [45] proposed a novel system involving automated tagging, recurrent neural network and state selection for detection. Researchers also developed a massive neural network to identify patterns in satellite data. There are also disadvantages and drawbacks associated with this version of Action Plan are global target recognition benchmarks. Readers are encouraged to turn to for more information about how to locate objects through remote sensing applications. The mission of licence plate detector is to identify vehicles speeding drivers. You will find more information in [30] if you wish. Anomaly identification can be used in a wide variety of uses such as fraud detection, temperature control and healthcare. Many existing anomaly detection methods analyzed a particular data point at a time. Since all the analyses agree, Barz et al. (2014) indicate major anomalous regions (anomalies) in the results.

(2) In military sector remote sensing objects detection is representative application: Remote sensing object recognition is a technology used to identify objects found in remote sensing photographs or recordings. First of all, the excessive large input number, however, limited targets make current object detection both too slow and too difficult to do. It causes plenty of false positives, causing both chaos and inconvenience. Researchers use the approach of data fusion to resolve the problems of distance learning. They were too eager in detecting small targets which contributed to so much inaccuracy that was reported. Feature engineering will create problems that current data domain transmission methods, such as Faster R-CNN, FCN, SSD and YOLO, will not be able to solve. It is difficult to develop various computer programs.

(3) An autonomous vehicle requires accurate vision of surroundings to run: An AV usually controls what it perceives through machine learning (or deep learning). There is a strong connection between visual awareness and object detection. 3D object detection approaches require three dimensions that show more precise object's size and position information. The first problem of monocular methods is the limited precision of the bounding boxes on the raw image. Secondly,

point cloud-based approaches project an image directly onto a voxel system without information loss. By merging images and point cloud to produce detectors, it is means of getting more reliable results more quickly. Recently, Lu et al. use a novel architecture that includes 3D convolutions and RNNs to achieve centimetre-level localization. Song et al. [178] released the standard benchmark of a 3D modelled autonomous driving vehicle. In order to achieve improved functions, Banerjee et al. [79] use a fusion algorithm. Please see a recent survey [80] about the problem. Both unmanned vehicles also autonomous heavy schemes are able to recognize road signs. For the sake of protection and the assurance of obeying the road laws, an effective real-time traffic sign detection system helps drivers acquire knowledge about the location of the future signals. Machine learning approaches solve this issue with higher efficiency.

(4) The medical profession uses medical imaging as a supplement to medical care: A computer-assisted diagnostic framework can help doctors recognize a wide variety of cancer. The key measures in a CAD system can be understood as being divided into image segmentation, feature extraction, classification and object detection. Therefore, there is usually data synthesis discrepancy between the source and destination fields. A domain adaptation approach is required to correctly identify medical photographs. Li et al. integrated CNN in glaucoma diagnosis to comprise a large-scale dataset. In busy scenes like store shelf displays, Goldman and her team suggested a novel precise object detector and Goldman provided a new item activity dataset for them. This technology is capable of identifying events occurring in real-world locations. Data quality is more complex today than ever with the advent of new actors. MED is capable of identifying different kinds of accidents. Yang et al. [10] developed an event classification scheme to analyze multi-domain research. Wang et al. [9] construct affinity graphs to establish social relationships among users for an event prediction mission. Schinas and colleagues [11] have established a way to detect accidents in a huge amount of data. Refer to the latest survey [12] for more detail. Pattern detection poses various challenges such as scene occlusion, variable lighting, posture differences and sensor noise. They also created solid foundation for future important works in both 2D images and 3D points. Your speech patterns are very dramatic. Caption creation means that computers create a caption for a given image. The main point is not to catch any of the semantics for pictures. This is a very challenging job in connecting computer vision with natural language processing.

Conclusion

Researchers have taken notice of object detection because of the advancement of deep learning, which serves as motivation for them to increase their performance on computer vision tasks. Despite the fact that deep detection models were created to address issues with general object detection, they are nevertheless used at a specific level to ensure tiny object identification is successful. Aiming for outstanding accomplishments in popular and state-of-the-art deep models as we evaluate tiny object detection using deep models, our objective is to present a wide range of perspectives on the use of deep models in object detection. Deep learning-based target identification technology has grown exponentially thanks to advances in computer science. One-stage detection models and two-stage detection models are the most common types of detection models. In comparison to another technique, models in the one-stage approach are referred to as detectors since they have better and more efficient detection. With this efficiency, real-time operations are possible, as well as practical applications. However, finding the right balance among speed and accuracy is a challenging problem that must be considered. Models in the two-stage technique, on the other hand, have a reputation as region-based detectors that are very accurate but move too slowly to be useful in the actual world. This problem arises as a result of the way networks are computed. High precision viable answers are more critical now than ever before if we are to reap the benefits of more precise implementations. Researchers know that the end goal is to use better detectors and processing equipment. Object detection is advancing quickly in response to the demands of the medical, military, transportation and daily life industries. The detecting industry has a lot of room for growth. This area's growth necessitates thought and cares going forward.

Despite recent successes, detection performance has increased dramatically, but the accuracy difference between conventional items and tiny objects remains enormous. The disparity between small, medium and large scales is too great in the COCO dataset's criteria. When used to identify little items, most models aren't very successful at detecting regular objects. Consequently, the first step in closing the detection gap for tiny objects is to invest in datasets with a large amount of data for training models and a wide variety of categories to compete with the human visual system. One-stage and two-stage detection models have been the most common so far in detection research it is called detectors when used to one-stage models because it provides superior detection efficiency than another strategy. Efficiency here has the ability to run in real-time and can be used to real-world situations. Accuracy and speed are challenging trade-offs and they must be considered in order to balance the gap. Models in the two-stage technique, on the other hand, have a reputation as region-based detectors that are very accurate but move too slowly to be useful in the actual world. This flaw can be traced back to how networks are computed. Based on our research, it's clear that architectures used as the foundation for extracting deep features have a big impact on frameworks. The better the detection accuracy, the more complex the architecture is. The more layers a network has, the more complex it becomes to train because of the massive number of parameters that must be learned. As a result, a large amount of data is required in order to fine-tune these parameters appropriately. Increases in computation will result in increased resource consumption. As a result, putting them to use in real-world scenarios will be tough. In addition, contextual usage in models is severely constrained, resulting in the omission of a significant amount of important and informative training data, particularly for tiny objects.

References

1. Dollar, P., Wojek, C., Schiele, B., Perona, P.: Pedestrian detection: an evaluation of the state of the art. IEEE Trans. Pattern Anal. Mach. Intell. **34**, 743–761 (2012)
2. Geiger, A., Lenz, P., Urtasun, R.: Are we ready for autonomous driving? the kitti vision benchmark suite. In: 2012 IEEE Conference on Computer Vision and Pattern Recognition, pp. 3354–3361, June 2012
3. Russakovsky, O., Deng, J., Su, H., Krause, J., Satheesh, S., Ma, S., Huang, Z., Karpathy, A., Khosla, A., Bernstein, M., Berg, A.C., Fei-Fei, L.: Imagenet large scale visual recognition challenge. Int. J. Comput. Vision **115**, 211–252 (2015)
4. Zhu, P., Wen, L., Bian, X., Ling, H., Hu, Q.: Vision meets drones: a challenge. arXiv preprint (2018). arXiv:1804.07437
5. Ren, S., He, K., Girshick, R., Sun, J.: Faster r-cnn: towards real-time object detection with region proposal networks. IEEE Trans. Pattern Anal. Mach. Intell. **39**, 1137–1149 (2017)
6. Redmon, J., Divvala, S., Girshick, R., Farhadi, A.: You only look once: unified, real-time object detection. In: 2016 IEEE Conference on Computer Vision and Pattern Recognition (CVPR), pp. 779–788, June 2016
7. Liu, W., Anguelov, D., Erhan, D., Szegedy, C., Reed, S., Fu, C.-Y., Berg, A.C.: Ssd: single shot multibox detector. In: Leibe, B., Matas, J., Sebe, N., Welling, M. (eds.) Computer Vision—ECCV 2016, pp. 21–37. Springer International Publishing, Cham (2016)
8. He, K., Gkioxari, G., Dollr, P., Girshick, R.: Mask r-cnn. In: 2017 IEEE International Conference on Computer Vision (ICCV), pp. 2980–2988, Oct 2017
9. Khan, A., Sohail, A., Zahoora, U., Qureshi, A.S.: A survey of the recent architectures of deep convolutional neural networks. arXiv preprint (2019). arXiv:1901.06032
10. Lin, T., Dollr, P., Girshick, R., He, K., Hariharan, B., Belongie, S.: Feature pyramid networks for object detection. In: 2017 IEEE Conference on Computer Vision and Pattern Recognition (CVPR), pp. 936–944, July 2017
11. Xie, S., Girshick, R., Dollar, P., Tu, Z., He, K.: Aggregated residual transformations for deep neural networks. In: Proceedings of the IEEE Conference on Computer Vision and Pattern Recognition, pp. 1492–1500 (2017)
12. Ghiasi, G., Lin, T.-Y., Pang, R., Le, Q.V.: Nas-fpn: learning scalable feature pyramid architecture for object detection. arXiv preprint (2019). arXiv:1904.07392
13. Howard, A.G., Zhu, M., Chen, B., Kalenichenko, D., Wang, W., Weyand, T., Andreetto, M., Adam, H.: Mobilenets: efficient convolutional neural networks for mobile vision applications. arXiv preprint (2017). arXiv:1704.04861
14. Zhang, X., Zhou, X., Lin, M., Sun, J.: Shufflenet: an extremely efficient convolutional neural network for mobile devices. In: Proceedings of the IEEE Conference on Computer Vision and Pattern Recognition, pp. 6848–6856 (2018)
15. Iandola, F.N., Han, S., Moskewicz, M.W., Ashraf, K., Dally, W.J., Keutzer, K.: Squeezenet: Alexnet-level accuracy with 50x fewer parameters and¡ 0.5 mb model size. arXiv preprint (2016) arXiv:1602.07360
16. Chollet, F.: Xception: deep learning with depth wise separable convolutions. In: Proceedings of the IEEE Conference on Computer Vision and Pattern Recognition, pp. 1251–1258 (2017)
17. Sandler, M., Howard, A., Zhu, M., Zhmoginov, A., Chen, L.-C.: Mobilenetv2: inverted residuals and linear bottlenecks. In: Proceedings of the IEEE Conference on Computer Vision and Pattern Recognition, pp. 4510–4520 (2018)
18. Wang, R.J., Li, X., Ling, C.X.: Pelee: a real-time object detection system on mobile devices. Adv. Neural Inform. Process. Syst. **24**, 1963–1972 (2018)
19. He, K., Zhang, X., Ren, S., Sun, J.: Deep residual learning for image recognition. In: 2016 IEEE Conference on Computer Vision and Pattern Recognition (CVPR), pp. 770–778, June 2016
20. Girshick, R., Donahue, J., Darrell, T., Malik, J.: Rich feature hierarchies for accurate object detection and semantic segmentation. In: 2014 IEEE Conference on Computer Vision and Pattern Recognition, pp. 580–587, June 2014

21. Girshick, R.: Fast r-cnn. In: 2015 IEEE International Conference on Computer Vision (ICCV), pp. 1440–1448, Dec 2015
22. Redmon, J., Farhadi, A.: Yolo9000: better, faster, stronger. In: 2017 IEEE Conference on Computer Vision and Pattern Recognition (CVPR), pp. 6517–6525, July 2017
23. Redmon, J., Farhadi, A.: Yolov3: an incremental improvement. CoRR, vol. abs/1804.02767 (2018)
24. Lin, T., Goyal, P., Girshick, R., He, K., Dollr, P.: Focal loss for dense object detection. In: 2017 IEEE International Conference on Computer Vision (ICCV), pp. 2999–3007, Oct 2017
25. Fu, C.-Y., Liu, W., Ranga, A., Tyagi, A., Berg, A.C.: Dssd: deconvolutional single shot detector. arXiv preprint (2017). arXiv:1701.06659
26. Zhao, Q., Sheng, T., Wang, Y., Tang, Z., Chen, Y., Cai, L., Ling, H.: M2det: a single-shot object detector based on multi-level feature pyramid network. arXiv preprint (2018). arXiv:1811.04533
27. Zhang, S., Wen, L., Bian, X., Lei, Z., Li, S.Z.: Single-shot refinement neural network for object detection. In: Proceedings of the IEEE Conference on Computer Vision and Pattern Recognition, pp. 4203–4212 (2018)
28. Hu, H., Gu, J., Zhang, Z., Dai, J., Wei, Y.: Relation networks for object detection. In: Proceedings of the IEEE Conference on Computer Vision and Pattern Recognition, pp. 3588–3597 (2018)
29. Dai, J., Qi, H., Xiong, Y., Li, Y., Zhang, G., Hu, H., Wei, Y.: Deformable convolutional networks. In: Proceedings of the IEEE International Conference on Computer Vision, pp. 764–773 (2017)
30. Zhu, X., Hu, H., Lin, S., Dai, J.: Deformable convnets v2: more deformable, better results. arXiv preprint (2018) arXiv:1811.11168
31. Shrivastava, A., Gupta, A., Girshick, R.: Training region-based object detectors with online hard example mining. In: Proceedings of the IEEE Conference on Computer Vision and Pattern Recognition, pp. 761–769 (2016)
32. Bell, S., Lawrence Zitnick, C., Bala, K., Girshick, R.: Inside-outside net: detecting objects in context with skip pooling and recurrent neural networks. In: Proceedings of the IEEE Conference on Computer Vision and Pattern Recognition, pp. 2874–2883 (2016)
33. Dai, J., Li, Y., He, K., Sun, J.: R-fcn: object detection via region based fully convolutional networks. In: Advances in Neural Information Processing Systems, pp. 379–387 (2016)
34. Zhu, Y., Zhao, C., Wang, J., Zhao, X., Wu, Y., Lu, H.: Couplenet: coupling global structure with local parts for object detection. In: Proceedings of the IEEE International Conference on Computer Vision, pp. 4126–4134 (2017)
35. Huang, J., Rathod, V., Sun, C., Zhu, M., Korattikara, A., Fathi, A., Fischer, I., Wojna, Z., Song, Y., Guadarrama, S., et al.: Speed/accuracy trade-offs for modern convolutional object detectors. In: Proceedings of the IEEE Conference on Computer Vision and Pattern Recognition, pp. 7310–7311 (2017)
36. Shrivastava, A., Sukthankar, R., Malik, J., Gupta, A.: Beyond skip connections: top-down modulation for object detection. arXiv preprint (2016). arXiv:1612.06851
37. Bodla, N., Singh, B., Chellappa, R., Davis, L.S.: Soft-NMS–improving object detection with one line of code. In: Proceedings of the IEEE International Conference on Computer Vision, pp. 5561–5569 (2017)
38. Cai, Z., Vasconcelos, N.: Cascade r-cnn: delving into high quality object detection. In: Proceedings of the IEEE Conference on Computer Vision and Pattern Recognition, pp. 6154–6162 (2018)
39. Singh, B., Davis, L.S.: An analysis of scale invariance in object detection snip. In: Proceedings of the IEEE conference on computer vision and pattern recognition, pp. 3578–3587 (2018)
40. Tychsen-Smith, L., Petersson, L.: Improving object localization with fitness nms and bounded iou loss. In: Proceedings of the IEEE Conference on Computer Vision and Pattern Recognition, pp. 6877–6885 (2018)
41. Kong, T., Sun, F., Yao, A., Liu, H., Lu, M., Chen, Y.: Ron: reverse connection with objectness prior networks for object detection. In: Proceedings of the IEEE Conference on Computer Vision and Pattern Recognition, pp. 5936–5944 (2017)

42. Law, H., Deng, J.: Cornernet: detecting objects as paired keypoints. In: Proceedings of the European Conference on Computer Vision (ECCV), pp. 734–750 (2018)
43. Braun, M., Krebs, S., Flohr, F., Gavrila, D.: Eurocity persons: a novel benchmark for person detection in traffic scenes. IEEE Trans. Pattern Anal. Mach. Intell. 1–1 (2019)
44. Zhang, S., Benenson, R., Schiele, B.: Citypersons: a diverse dataset for pedestrian detection. In: Proceedings of the IEEE Conference on Computer Vision and Pattern Recognition, pp. 3213–3221 (2017)
45. Li, X., Flohr, F., Yang, Y., Xiong, H., Braun, M., Pan, S., Li, K., Gavrila, D.M.: A new benchmark for vision-based cyclist detection. In: 2016 IEEE Intelligent Vehicles Symposium (IV), pp. 1028–1033. IEEE (2016)
46. Dalal, N., Triggs, B.: Histograms of oriented gradients for human detection. In: international Conference on computer vision and Pattern Recognition (CVPR'05), vol. 1, pp. 886–893. IEEE Computer Society (2005)
47. Ess, A., Leibe, B., Van Gool, L.: Depth and appearance for mobile scene analysis. In: 2007 IEEE 11th International Conference on Computer Vision, pp. 1–8. IEEE (2007)
48. Wojek, C., Walk, S., Schiele, B.: Multi-cue onboard pedestrian detection. In: 2009 IEEE Conference on Computer Vision and Pattern Recognition, pp. 794–801. IEEE (2009)
49. Enzweiler, M., Gavrila, D.M.: Monocular pedestrian detection: survey and experiments. IEEE Trans. Pattern Anal. Mach. Intell. **31**(12), 2179–2195 (2008)
50. Zoph, B., Cubuk, E.D., Ghiasi, G., Lin, T., Shlens, J., Le, Q.V.: Learning data augmentation strategies for object detection. CoRR, vol. abs/1906.11172 (2019)
51. Kong, T., Sun, F., Liu, H., Jiang, Y., Shi, J.: Fovea box: beyond anchor-based object detector. arXiv preprint (2019). arXiv:1904.03797
52. Pang, J., Chen, K., Shi, J., Feng, H., Ouyang, W., Lin, D.: Librar-cnn: Towards balanced learning for object detection. arXiv preprint (2019). arXiv:1904.02701
53. Kim, S.-W., Kook, H.-K., Sun, J.-Y., Kang, M.-C., Ko, S.-J.: Parallel feature pyramid network for object detection. In: Proceedings of the European Conference on Computer Vision (ECCV), pp. 234–250 (2018)
54. Li, Z., Zhou, F.: Fssd: feature fusion single shot multibox detector. arXiv preprint (2017). arXiv:1712.00960
55. Chen, Y., Li, J., Zhou, B., Feng, J., Yan, S.: Weaving multi-scale context for single shot detector. arXiv preprint (2017). arXiv:1712.03149
56. Zheng, L., Fu, C., Zhao, Y.: Extend the shallow part of single shot multibox detector via convolutional neural network. In: Tenth International Conference on Digital Image Processing (ICDIP 2018), vol. 10806, p. 1080613. International Society for Optics and Photonics (2018)
57. Bae, S.-H.: Object detection based on region decomposition and assembly. arXiv preprint (2019). arXiv:1901.08225
58. Liu, Y., Wang, R., Shan, S., Chen, X.: Structure inference net: object detection using scene-level context and instance-level relationships. In: Proceedings of the IEEE Conference on Computer Vision and Pattern Recognition, pp. 6985–6994 (2018)
59. Singh, B., Najibi, M., Davis, L.S.: Sniper: efficient multi-scale training. In: Advances in Neural Information Processing Systems, pp. 9310–9320 (2018)
60. Barnea, E., Ben-Shahar, O.: On the utility of context (or the lack thereof) for object detection. CoRR, vol. abs/1711.05471 (2017)
61. Liang, K., Chang, H., Ma, B., Shan, S., Chen, X.: Unifying visual attribute learning with object recognition in a multiplicative framework. IEEE Trans. Pattern Anal. Mach. Intell. **41**, 1747–1760 (2019)

Chapter 8
Routing Protocols: A Survey

K. Raghavendra Rao and B. N. Jagadesh

Introduction

Nodes in the network can share information from one to another, in MANET network, nodes can share information without using any existing infrastructure, and it can share packets from one node to another node. The packet finally can reach the destination node with the help of intermediate nodes [1]. Routing is basic step for transmission of data. Most of the researchers put their efforts in finding the optimal path. Routing can be very sensitive in terms of different characteristics.

Throughput: Throughput is nothing but how fast we can transfer the data from source to destination. It means it is actual measure of how fast we can transfer the data.

Latency: Latency or delay can define how long it will take to reach data from source to destination completely.

Bandwidth: It is the capacity of the link that can transfer the data from one node to another node within a given amount of time. It is one of the main characteristics of network to measure the performance of that network.

Queuing time: It is nothing but a node can hold a message before it can process.

Energy: Nodes in the network can rely on cooperation. But selfish node can preserve their resource for their purpose. But altruistic node in the network can spend some energy to transfer.

K. Raghavendra Rao (✉)
Koneru Lakshmaiah Education Foundation, Vijayawada, Andhra Pradesh, India
e-mail: raghava514.k@gmail.com

B. N. Jagadesh
School of Computer Science and Engineering, VIT-AP University, Amaravathi, Andhra Pradesh, India

Ch. Satyanarayana et al. (eds.), *Proceedings of the International Conference on Computer Vision, High Performance Computing, Smart Devices and Networks*, Advanced Technologies and Societal Change,
https://doi.org/10.1007/978-981-19-4044-6_9

Load: Another important characteristic of network is load. If a network can face a huge amount of data, then automatically network performance can be reduced.

Congestion: Congestion in the network can occur, when different nodes can transfer the data from source to destination. But all nodes have same intermediate node. If the node can face congestion, means the no of packets that node can receives is greater than it's capacity.

Efficiency: It is nothing but how efficiently it can exchange information.

Shortest path: Path is nothing but in which direction you can deliver the data. In network, we can select the minimum path from source to destination to deliver the data.

Multiple paths: When we are delivering the data from source to destination, we can select multiple paths from source to destinations to deliver the data in very fast.

Mobility: Mobility means node can change its position in the network frequently. Node position can also affect the network performance.

Security: It is another characteristic of network; node can deliver the data in very secure manner which means that we can protect out data from eavesdrops.

Attack: In the network, some malicious node can attack another node, to stop data transfer in the network.

Packet loss: When node can face congestion, mobility of next node, no link to next node any one of the mentioned characteristic, it can simply drop the packet.

Routing Protocols and Classification

Routing Protocols

Routing in adhoc network can be broadly classified as topology-based or position-based approach. In this paper, we are mainly concentrated on topology-based routing. Topology base routing can depend on existing information about the network. It can further subdivided into three types, namely (A) Proactive (B) Reactive (C) Hybrid routing.

These three categories are again classified into subcategories.

Proactive routing can also called table-driven protocol. Reactive routing can also be called on-demand routing protocol, and hybrid is a combination of both proactive and reactive routing.

These protocols are used to transfer the data from source node to destination node without using any infrastructure which means that it can be self-configure network (Fig. 8.1).

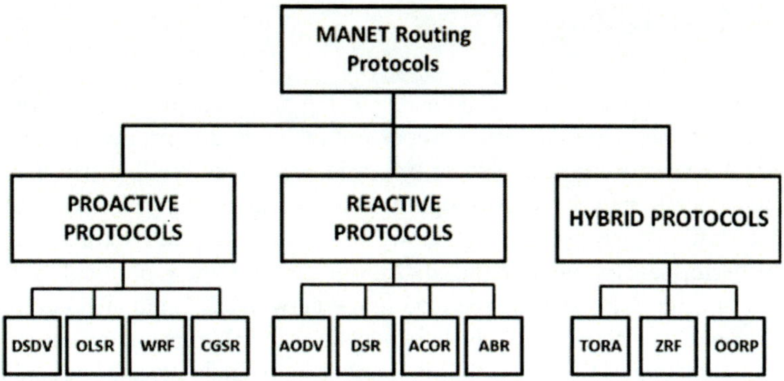

Fig. 8.1 Different types of routing protocols in MANET

Proactive Routing Protocol

Kim et al. [2] proposed a new algorithm called modified-TCP-Vegas for adhoc networks because wired and wireless RTT is different. Mnaouer et al. [3], he developed a new protocol for data delivery, combining the power awareness and adaptability. He improved the efficiency and reduced response time by using optimize hybrid multicast routing. Roy et al. [4], he proposed new routing technique for multimedia data. He can utilize the ACO framework to achieve his goal. Here, node can select stable route from available routes. He combines the ACO and OLSR for stable routes. Safa et al. [5], In manet nodes are equipped with different transmission technologies like Wi-Fi and Bluetooth. Nodes in the network are needed to support interoperability between heterogeneous nodes. If node want to communicate to another node, then then major constrain is power. He proposed power aware routing protocol. It can check the power of the node before entering the route in the table. Semchedine and Moussaoui [6], in any network data can be protected with protocol. Because the security depends on using protocol. He proposed new protocol CRY-OLSR to protect data in network. Nithya [7], in MANET by using clustering data can transfer groups simultaneously. Clustering can remove scalability problem in network. He proposed new protocol PSR for this purpose. Gadekar [8], in networks, we can protect node from attacks. One of the attacks is node isolation attack. We can detect and mitigate this by suing Modified-OLSR. Shakkeera [9], nodes can transfer the data from one node to another node. We are using one of the proactive routing algorithms, in that we can make constrain how many MPR can select. Necessity first algorithm can be used for this purpose. Zhou [10], MANET node wants to transfer the data to another node. It can select the path to reach the destination. Here, we are using OLSR. In data transfer path selection can play a key role. He proposed AOMDV for selecting first path considering the state of that node. It means that we can select multiple nodes in our route, and the algorithm can decide based on the node weight it is in route or not.

Reactive Routing Protocol

Reactive routing is also called on-demand routing protocols. It can create the route on demand, means when required. It does not maintain any data related to routes of different paths. It can create route when node requires sending data from one to another node. Zhang [11] has proposed load balancing algorithm for MANET, to reduce traffic congestion, and generally when load increases automatically, the network performance can be reduced. Author develops new protocol called ant colony optimization for sharing the load in the network. Krishna [12] developed a new approach for selecting multiple disjoint paths using Qos ant colony optimization method. Al-Rabayah [13] proposed topology-based vehicle monitoring and traffic monitoring system. This can solve the routing overhead problem with the help of reactive and geo routing. It will take time to again create the route. Valera et al. [14, 15] proposed a novel algorithm for multiple paths creation. It can create the multiple shortest paths to destination. It can also specify to make network robust. It can improve the packet delivery decrease delay and reduce overhead routing with the help of small information in cache. Agglou and Tafazolli [16], routes can be repaired and discovered. We can search route with the help of route search algorithm. He mainly concentrated on route repair mechanism. It can discover and repair the route. Lin et al. [17] proposed different wireless technologies for better, downloading data. It means that he can create a framework for better communication of different networks. Wu and Harms [18], selecting multiple paths, generally data cannot ravel in one path, and source can have secondary path for transfer the data when path one can fail. But Kui proposed multipath for transfer the data. LWPSR [19] has proposed by Zehua, and he mainly concentrates on new routing protocol. Existing protocols are not maintaining more topological information for forwarding opportunistic data. So, the LWPSR can be used for this purpose. Protocols have proposed for transferring the data from node to node. But no protocol can take mobility as a characteristic. Feng et al. [20] have taken a mobility as consideration for transferring the data. He calculated the speed of source and destination to transfer the data. Xu et al. [21], in this, it reduces the power consumptions with the help of multinode cooperation to deliver the data. The proposed bandwidth PAMP routing. It can send data in multiple paths. Seo and Thottethodi [22], he says that streaming application is very important. He proposes the novel communication for streaming data. Flexible streaming can be expressed, and node disjoint path can reduce the link contention problem and maximizes the bandwidth. Papapetrou [23], all routing protocols are concentrating on route stability. But no one considers the connection. He proposed the novel protocol for connection stability. Sarkar [24], selection of path like optimal is critical in MANET. He combines the features of AODV and AOC to improve the quality of service. Based on the new protocol, he can select best path from source to destination. Rodrı´guez [25], generally nodes can have capability to communicate each other. MANET has proposed to communicate nodes when they are not in near to each other. Juan proposed the protocol for communicating collaboratively in network. The protocol works in high-level layer. Name of the protocol is HLMP.

Gaurav [26], stability of MANET can be achieved with the help of virtual backbone. Here, only particular node can maintain the topological information. Only those nodes can participate in routing of messages. He proposed the security in virtual backbone. He can secure the network with the help of distributed information. Hussain [27], in network we can find the path with the help of route request broadcast message. It will create little traffic in certain conditions. He can propose the novel technique to broadcast the packet. Nodes can be cluster. The cluster node can take the responsibility of broadcasting the packet. Safa [5], In manet nodes are equipped with heterogeneous technologies like Wi-Fi and Bluetooth. Nodes in the network are needed to support interoperability between heterogeneous nodes. If node want to communicate to another node, here Major constrain is power. He proposed power-aware routing protocol. It can check the power of the node before entering the route in the table. Malathi [28], nodes can transfer the data without using any network if those are in near range, if not it can use any network to transfer the data. During transmission, sometime link can fail. He proposed new protocol for making reliable route of considering different parameters before taking link. Those can make connection reliable during transmission, link quality, capacity of channel and node energy. For this, he proposed P2R2 protocol. Lee [29], most of the protocol can create single route from source to destination. If path can broke, protocol can again perform the route request process. By using SMR, we can create and use multiple paths for data transmission. Vahedi et al. [30], proposed a protocol, it can create multiple paths from source to destination. In this, each data transfer can have three paths. One is shortest delay, and two are maximally disjoint routes. Zafar [31], data can travel in node to node. If link can break in between source to destination, this protocol can select another path, in that path not include the broken link node. He can propose the PMPR protocol. Bai [32] proposed a new routing protocol for handling of RREP. It can handle in two ways, one is to look its cache for path, second is to perform one hop SRR. Kajikawa [33], proposed a new protocol, it is a combination of two protocols. By using this protocol, we can construct least hop backup path and reduce packet drops using salvaging method. Medidi [34] proposed novel approach for selecting the path. He can monitor the all routes based on the history of node. Like how many packet it forwarded and how many drop. Based on this history, a node can select, to keep that node in the path or not. Marti et al. [35] proposed new techniques to identify the misbehavior nodes in the network. He proposed two concepts; one is watch dog, and another one is path rater. By combining of these two, we can find that the node is good or misbehavior node. Chang [36] proposed a new incentive mechanism, for selfish node. He provides the incentives for node, when node can transfer the data.

Hybrid Routing Protocol

It is a combination of reactive and proactive routing. VANET is mobile technology that is communicating of vehicles. For security and communication purpose, Fazio [37] proposed a cross-layer routing in VANET. Fazio proposed a Position Based

Routing Protocol. Position-based routing means that it can determine the position of node before delivery the data. A location service can take the responsibility to identify the node location. In network, data can forward the node with the help of information from packet header. If information is not available, the node can see his cache to forward the data. To protect the cache content, Kiskani and Sadjadpour [38] proposed a new coded technique for storing content n cache. He can make normal data as coded data using XOR operation. Yang [39], he proposed, data can be delivered, reliable and timely manner in dynamic nature of node. He can transfer the data based on the position of node. Ata can be deliver the source the intermediate nodes can deliver the data, here the forwarding nodes can cache the packet, if one of the candidate can fail to forward the next node can deliver the data.

Nzouonta et al. [40] proposed road-based vehicle routing. Existing protocol can perform the operation using city-based. It uses the position of destination for forwarding the data. It can use the real-time traffic for data forwarding. Kajikawa et al. [33], proposed a new protocol, it is a combination of two protocols. By using this protocol, we can construct least hop backup path and reduce packet drops using salvaging method.

Other Routing Protocols

Other routing protocols are mainly classified based on their behavior of routing. These have mainly taken different metrics to forward the data from node to node. Yu et al. [41] proposed a more aggregate throughput by accepting concurrent communication. Cam not allows aggressive communication. He can perform concurrent communication with the help of packet salvaging and less sensing protocol. Rios [42] proposed a new protocol for communication between vehicle and road side unit. He has taken density also for consideration. Shabut et al. [43] can propose the trust and energy-aware link from source to destination. He can propose trust management-QoS framework for data delivery. Rajeswari and Ravi [44], proposed a new novel algorithm for scalability and heterogeneity in MANET. The main aim of this protocol is to increase the performance of network and resource management. Biradar [45], If we want to communicate multiple people, then we can establish a connection or creation of link is required. The author proposed to select the reliable neighbor using MMRNS. Papadimitratos [46], in MANET data can be transferred from one node to another node. Here, data can travel in wireless medium. So, eavesdroppers are trying to access the data. We can protect our data. He can propose new protocol for protecting the data. The protocol is SMT. Zhang [47] provides a geo casting delivery. He mainly concentrates on mobility. He proposed that a new algorithm is node mobility aware and contact-aware geo casting.

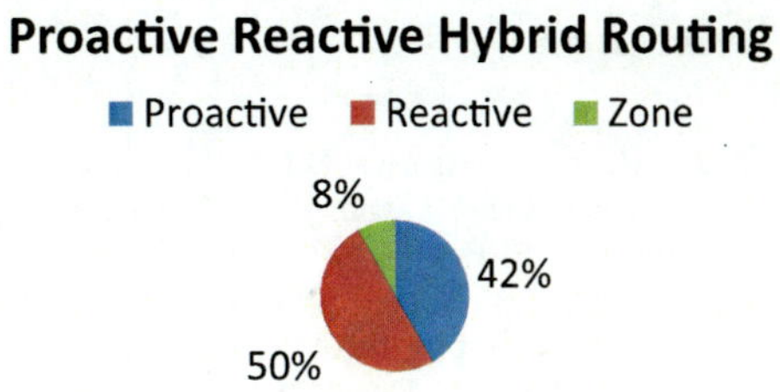

Fig. 8.2 Represent the utilization of routing protocol

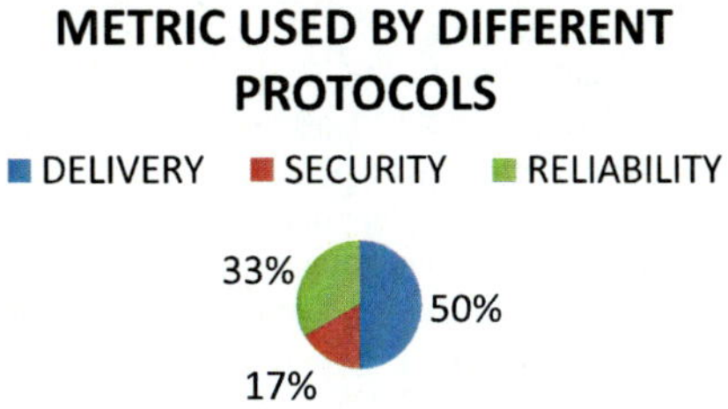

Fig. 8.3 Metric used by different authors

Methods Utilization Chart

i) The below chat explored that the authors are contributed in different methods of routing. By looking below chart, we can understand that most of the authors have used reactive routing (Fig. 8.2).

ii) Metric utilization chart: The below chart explains that various authors are concentrated metrics. Majority of the authors are mainly concentrated on packet delivery. Like they are giving more preference to delivery. Once data delivery is fast, we can send more number of packets. And reliability is also majorly concentrated on after delivery which means that authors are giving preference either fastest delivery or reliability (Fig. 9.3).

Conclusion

The paper explored various methods of routing, and various techniques are performed in MANET. This paper focused on routing techniques like proactive, reactive and zone routing. Like authors are concentrated on different metrics like delay, performance, optimal path, energy consumption, bandwidth, mobility, congestion, etc. But more authors are concentrated on selecting shortest path. Some are concentrated on reliability of path, security of the link, etc. We are identified that there is lot of research needed to focus on identifying multiple paths to transfer the data form source to destination at the same time.

References

1. De Morais, C., Agarwal D.P.: Adhoc & sensor networks theory and applications
2. Kim, D., Bae, H., Toh, C.K.: Improving TCP-Vegas performance over MANET routing protocols. IEEE (2007)
3. Mnaouer, A.B., Chen, L., Foh, C.H., Tantra, J.W.: OPHMR: an optimized polymorphic hybrid multicast routing protocol for MANET. IEEE Trans. Mobile Comput. **6**(5) (2007)
4. Roy, B., Banik, S., Dey, P., Sanyal, S., Chaki, N.: Ant colony based routing for mobile ad-hoc networks towards improved quality of services. J. Emerg. Trends Comput. Inf. Sci. **3** (2012)
5. Safa, H., et al.: PHAODV: power aware heterogeneous routing protocol for MANETS. J. Network Comput. Appl. (2014). https://doi.org/10.1016/J.Jnca.2014.07.035i
6. Semchedine, F., Moussaoui, A.: CRY OLSR: crypto optimized link state routing for MANET. IEEE (2016). 978-1-5090-5146-5/16
7. Nithya, S., Kumar G.A., Adhavan P.: Destination-sequenced distance vector routing (DSDV) using clustering approach in mobile ad hoc network. In: 2012 International Conference On Radar, Communication And Computing (ICRCC)
8. Gadekar, S., Kadam, S.: Secure optimized link state routing (OLSR) protocol against node isolation attack. In: IEEE International Conference on Power, Control, Signals And Instrumentation Engineering (ICPCSI-2017)
9. Shakkeera.: Optimal path selection technique for flooding in link state routing protocol using forwarding mechanisms in MANET. In: Proceedings of the International Conference on Communication and Computational Intelligence—2010
10. Zhou, J., Xu, H., Qin, Z., Peng, Y., Lei, C.: Ad hoc on-demand multipath distance vector routing protocol based on node state. Commun. Network, **5**, 408-413 (2013) https://doi.org/10.4236/Cn.2013.53B2075
11. Zhang, D., Zhou, D.: Load balancing algorithm based on history information in MANET. In: 2017 IEEE 2nd Information Technology, Networking, Electronic And Automation Control Conference (ITNEC) (2017)
12. Krishna, P.V.: Quality-of-service-enabled ant colony-based multipath routing for mobile ad hoc networks. IET Commun. **6**(1), 76–83 (2012). https://doi.org/10.1049/Iet-Com.2010.0763
13. Al-Rabayah, M., Malaney, R.: A new scalable hybrid routing protocol for VANETS. IEEE Trans. Veh. Technol. **61**(6) (2012)
14. Valera, A., Seah, W.K.G., Rao, S.V.: Cooperative packet caching and shortest multipath routing in mobile ad hoc networks. In: IEEE INFOCOM 2003 twenty-second annual joint conferences of the IEEE Computer and Communications Societies (IEEE Cat. No.03CH37428) (2003)
15. Valera, A.C., Seah, W.K.G., Rao, S.V.: Improving protocol robustness in ad hoc networks through cooperative packet caching and shortest multipath routing. IEEE Trans. Mobile Comput. **4**(5) (2005)
16. Agglou, G., Tafazolli, R.: Determining the optimal configuration for the relative distance microdiscovery ad hoc routing protocol. IEEE Trans. Veh. Technol. **51**(2) (2002)
17. Lin, T.Y., Huang, T.Y., Hsu, C.F.: Synergizing wireless communication technologies to improve internet downloading experiences. IEEE Trans. Comput. **63**(11) (2014)
18. Wu, K., Harms, J.: Multipath routing for mobile ad hoc networks. J. Commun. Networks **4**(1) (2002)
19. Wang, Z., Chen, Y., Li, C.: PSR: a lightweight proactive source routing protocol for mobile ad hoc networks. IEEE Trans. Veh. Technol. **63**(2) (2014)
20. Feng, K.T., Hsu, C.H., Lu, T.E.: Velocity-assisted predictive mobility and location-aware routing protocols for mobile *ad hoc* networks. IEEE Trans. Veh. Technol. **57**(1) (2008)
21. Xu, H., Huang, L., Qiao, C., Zhang, Y., Sun, Q.: Bandwidth-power aware cooperative multipath routing for wireless multimedia sensor networks. IEEE Trans. Wirel. Commun. **11**(4) (2012)
22. Seo, D., Thottethodi, M.: Disjoint-path routing: efficient communication for streaming applications. In: IEEE International Symposium On Parallel & Distributed Processing 2009 (2009)

23. Papapetrou, E., Pavlidou, F.-N.: Disjoint routes for on demand routing protocols in ad hoc networks. In: Globecom'01. IEEE Global Telecommunications Conference
24. Sarkar, D., et al.: Enhanced-Ant-AODV for optimal route selection in mobile ad-hoc network. J. King Saud Univ. Comput. Inf. Sci. (2018). https://doi.org/10.1016/J.Jksuci.2018.08.013
25. Rodríguez-Covili, J., Ochoa, S.F., Pino, J.A.: High level MANET protocol: enhancing the communication support for mobile collaborative work. J. Network Comput. Appl. **35**, 145–155 (2012)
26. Gaurav, A., Singh, A.K.: Light weight approach for secure backbone construction for MANETS. J. King Saud Univ. Comput. Inf. Sci. (2018). https://doi.org/10.1016/J.Jksuci.2018.05.013
27. Hussain, S.Z., Ahmad, N.: Minimizing broadcast expenses in clustered ad-hoc networks. J. King Saud Univ. Comput. Inf. Sci. **30**, 67–79 (2018)
28. Malathi, M., Jayashri, S.: Robust against route failure using power proficient reliable routing in MANET. https://doi.org/10.1016/J.Aej.2016.10.004
29. Lee, S.J., Gerla, M.: Split multipath routing with maximally disjoint paths in ad hoc networks. In: ICC 2001 IEEE International Conference on Communications. Conference Record (Cat. No.01CH37240)
30. Vahedi, S., Mohseni, M., Darehshoorzadeh, A.: Design a multi-path routing algorithm in ad hoc networks in order to improve fault tolerance. In: The 18th Annual IEEE International Symposium on Personal, Indoor and Mobile Radio Communications (PIMRC'07)
31. Zafar, H., Harle, D., Andonovic, I., Ashraf, M.: Partial-disjoint multipath routing for wireless ad-hoc networks. In: 32nd IEEE Conference on Local Computer Networks. IEEE, 0742-1303/07 2007. https://doi.org/10.1109/LCN.2007.84
32. Bai, R., Singhal, M.: Salvaging route reply for on-demand routing protocols in mobile ad-hoc networks. In: Mswim'05 Proceedings of the 8th ACM International Symposium on Modeling, Analysis and Simulation of Wireless and Mobile Systems
33. Kajikawa, H., Fukuhara, T., Bouk, S.H., Sasase, I.: Multipath routing protocol combined with least hop backup path and packet salvaging for MANETS. 978-1-4244-4561-5/09/2009 IEEE PACRIM'09
34. Medidi, S., Cappetto, P.: History-based route selection for reactive ad hoc routing protocols. 1-4244-1230-7/07/2007IEEE ICON
35. Marti, S., Giuli, T.J., Lai, K., Baker, M.: Mitigating routing misbehavior in mobile ad hoc networks. In: Proceeding Mobicom'00 Proceedings of the 6th Annual International Conference on Mobile Computing and Networking, pp. 255–262
36. Chang, J.M., Tsou, P.C., Woungang, I., Chao, H.C., Lai, C.F.: Defending against collaborative attacks by malicious nodes in MANETs: a cooperative bait detection approach. IEEE Syst. J.
37. Fazio, P., De Rango, F., Sottile, C.: A predictive cross-layered interference management in a multichannel MAC with reactive routing in VANET. IEEE Trans. Mobile Comput. **15**(8) (2016)
38. Kiskani, M.K., Sadjadpour, H.R.: A secure approach for caching contents in wireless ad hoc networks. IEEE Trans. Veh. Technol. **66**(11) (2017)
39. Yang, S., Yeo, C.K., Lee, B.S.: Toward reliable data delivery for highly dynamic mobile ad hoc networks. IEEE Trans. Mobile Comput. **11**(1) (2012)
40. Nzouonta, J., Rajgure, N., Wang, G., Borcea, C.: VANET routing on city roads using real-time vehicular traffic information. IEEE Trans. Veh. Technol. **58**(7) (2009)
41. Yu, C., Shin, K.G., Song, L.: Maximizing communication concurrency via link-layer packet salvaging in mobile ad hoc networks. IEEE Trans. Mobile Comput. **6**(4) (2007)
42. Rios, M.: Geopps-N: opportunistic routing for VANET in a public transit system. IEEE Lat. Am. Trans. **14**(4) (2016)
43. Shabut, A.M., Kaiser, M.S., Dahal, K.P., Chen, W.: A multidimensional trust evaluation model for MANETS. J. Network Comput. Appl. https://doi.org/10.1016/J.Jnca.2018.07.008
44. Rajeswari, P., Ravi, T.N.: He-series: an inventive communication model for data offloading in MANET. Egypt. Inform. J. **19**, 11–19 (2018)

45. Biradar, R.C., Manvi, S.S.: Neighbor supported reliable multipath multicast routing in MANETS, https://doi.org/10.1016/J.Jnca.2011.12.008
46. Papadimitratos, P., Haas, Z.J.: Secure message transmission in mobile ad hoc networks. Ad Hoc Networks **1**, 193–209 (2003)
47. Zhang, L., Yu, B., Pan, J.: Geomobcon: a mobility-contact-aware geocast scheme for Urban VANETS. IEEE Trans. Veh. Technol. **65**(8) (2016)

Chapter 9
Performance Evaluation of Segmentation Algorithms in Non Contrast and Contrast MRI Images for Region of Interest

S. Prabhu Das, B. N. Jagadesh, and B. Prabhakara Rao

Introduction

Biomedical Image Processing is the now becoming an emerging field in health monitoring, diagnosis and to treat different kinds of internal organ abnormalities or diseases for scientific study or treatment [1].

Medical Imaging [2] is an in vivo imaging process and it involves visualisation of internal organs within the body of humans or animals using non-invasive imaging modalities such as X-Ray, CT, MRI, PET Scans, etc. with the help of computer aids. Computer Aided Diagnosis (CAD) helps the medical expert to acquire better anatomical structures and to study the Region of Interest (ROI) within the tissue.

CAD tools involve the expert to automate the process [3], to get the results fast and accurately even for large number of cases without fatigue or data overload at minimal cost. They support remote accessing using information technologies through faster communication in case of emergency.

Imaging Modalities and Their Contrast Media

There are many imaging modalities [4] are used in biomedical engineering. They are Computed Tomography (CT), Magnetic Resonance Imaging (MRI), Ultrasound (US) Imaging, Digital Mammography, Positron Emission Tomography (PET) and

S. Prabhu Das (✉) · B. Prabhakara Rao
Department of Electronics and Communication Engineering, JNTUK, Kakinada, India
e-mail: spdas3607@gmail.com

B. N. Jagadesh
School of Computer Science and Engineering, VIT-AP University, Amaravathi, India

Ch. Satyanarayana et al. (eds.), *Proceedings of the International Conference on Computer Vision, High Performance Computing, Smart Devices and Networks*, Advanced Technologies and Societal Change,
https://doi.org/10.1007/978-981-19-4044-6_10

Thermal Imaging and so on. Each and every modality has its own strengths and weaknesses but they are application specific. For example, X-Ray Imaging is mainly used to visualise hard tissues such as bones within the body whereas CT and MRI imaging modalities are used to observe soft tissues or both. These processes include Ionisation and electromagnetic interferences at microscopic level, respectively.

CT: Computed Tomography (CT) as the name indicates that the computer creates the images sequence by taking the inputs from X-Rays at all possible angles around the object in the form of slices called Tomograms. CTs produce the pictorial information of the object using ionising radiation. It can visualise brightly the high-density matters such as bones, calcium and high densities and darkly the low-density matters such as liquids fats, etc. The Hounsfield units are used to represent the different densities measured. For example, Water typically having 0 HU may be varying between -7 to $+7$ HU, bones have higher densities greater than 500 HU, soft tissues within the range 10–60 HU and fats with low density between -25 HU to -250 [5].

Modern CT scanners are advantageous that they can scan the abdomen and pelvis very quickly at a slice thickness of less than 1 mm within the span of 6–7 s per view. This is better suitable for the case of children without giving anaesthesia. Detailed information about any tissue is also obtained by other projections such as Sagittal and Coronal. CT scanners can also produce 3D scans [5].

CT scanners produce high-resolution images with minimal noise but are unable to detect inflammation defects.

MRI: MRI [6] is a non-ionising modality in radio imaging and it is extensively used for imaging soft tissues, organs, bone and other internal structures because it offers much greater contrast between the diverse soft tissues of the body. It is mostly used in radiology to diagnose or monitor treatment for abnormal conditions or staging the disease within the chest, abdomen and pelvis. MRIs produce very good contrast images of soft tissues compared to CTs. The human body naturally consists of hydrogen atoms within the body. MRI consists of a powerful rotating magnet, electric field gradients [7] (induction coils) and a digital computer to form images. The protons within the hydrogen atoms of water molecules are having magnetic properties such that all protons are aligned parallel to the main magnetic field. In MRI, a strong magnetic field is applied to line up the nuclear magnetisation of hydrogen atoms of the observing part of the body. An RF signal from the induction coils is applied subsequently in the form of RF pulses to alter the alignment of the protons and reversed to get relaxation energy from the protons. This relaxation energy is detected by MRI scanners to construct the image using K-space [8]. By changing the parameters of this RF pulse sequence, different contrasts may be generated between tissues based on the relaxation properties of the hydrogen atoms there in the tissues.

Depending on relaxation times and dynamic contrast variations [9], the intensities within the image are correlating with the tissue characteristics. It can produce morphological and functional information about the tissue without any ionising radiation.

PET: Positron Emission Tomography (PET) [10] is a functional imaging technique used to obtain the information about the changes in metabolic process, oxygen through the blood flow and tissues or organs' working conditions. It is the combination of nuclear medicine and biochemical interactions with the body of observation. PET scans are mostly used for monitoring the functioning of brain disorders, heart problems and cancer cells within the body. PETs employ radioactive tracers that were swallowed as gas or injected into the body parts being examined. These tracers chemically react differently with the different parts of the body. These differential changes will be collected by the PET scanners as bright spots if they were infected. PET scanners can able to measure the oxygen [11] and sugar levels of the body and early stages of defects which are not possible by other scanners at cellular level. In PETs, gamma cameras are used to produce and to collect gamma rays as similar to X-rays. High chemical reaction of tracers will be shown as hot spots with greater intensity and less chemical reactions as cold spots with light intensities. Cancer cells [12] have high metabolic rate viewed as hot spots than non-cancer cells. Healthy tissues have more tracer absorption than the diseased one. Thus the level of heart problems was detected based the degree of interaction and colour differentiation. PET scanners are also used for detecting problems through the interaction of tracer with the glucose such as Alzheimer's disease [13], depression, epilepsy, head and Parkinson's disease in central nervous system. PETs also used with CT/MRI to get more information about disease.

PET scans are less accurate in case of small-sized tumours and at high blood sugar levels.

Microscopy [14] and PA imaging techniques [15] with contrasts are useful to detect defects from the defected organs due to deeper penetration.

According to our findings [16], ultrasonography examination could be used as a preliminary assessment for stratifying patients based on their risk of ADPKD progression. Because of the low accuracy and reproducibility of ultrasound, this estimate should only be used for patients with kidney volumes that are close to normal. A first magnetic resonance imaging (MRI) scan is recommended for dimensional increases in kidney size.

Contrast Materials: To improve medical imaging, various types of contrast media have been utilised. These contrast materials can be injected into veins or arteries, spinal discs or fluid spaces and other bodily cavities. Contrast materials come in a variety of forms:

1. X-ray and computed tomography (CT) imaging exams use iodine-based and barium-sulfate chemicals. The most often used contrast substance is barium-sulfate, which is administered orally. It is also administered rectally and comes in a variety of forms. Iodine-based and barium-sulfate contrast materials block or limit the ability of X-rays to pass through a specific area of the body. As a result, blood vessels, organs and other bodily tissue containing iodine-based or barium compounds on X-Ray or CT pictures change appearance.
2. Gadolinium is a key component of the most used magnetic resonance (MR) contrast material. When this chemical is present in the body, it changes the magnetic properties of adjacent water molecules, improving the quality of magnetic resonance imaging (MRI) images. Thus, internal soft organs, such

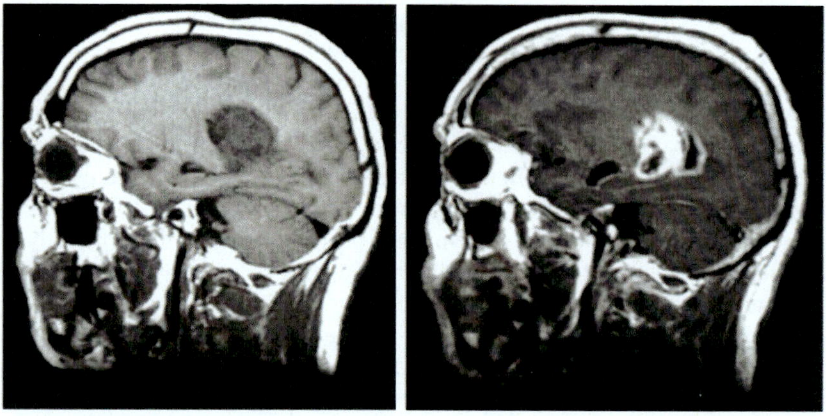

(a) MRI Image without Contrast (b)MRI Image without Contrast

Fig. 9.1 Illustration example for non-contrast and Gd Contrast MRI images. *Source* https://home.physics.wisc.edu/gilbert/wp-content/uploads/sites/3/2017/08/mri.gif

as the heart, lungs, liver, adrenal glands, kidneys, pancreas, gall bladder, spleen, uterus and bladder are monitored.

3. In imaging studies, saline (salt water) and gas (such as air) are also employed as contrast materials. Ultrasound imaging exams, particularly those of the heart, have used microbubbles and microspheres. Blood perfusion in organs is assessed using contrast-enhanced ultrasonography with microbubbles [17], thrombosis, such as in myocardial infarction, cardiac anomalies, liver and kidney tumours, inflammatory activity in inflammatory bowel illness and reaction to chemotherapy treatment (Fig. 9.1).

Proposed Methodology

In this paper, the proposed block diagram as shown in Fig. 9.2 is a new combination of denoising algorithms used to improve the image quality and to enhance the image contrast before segmenting the MRI Image. The image quality is improved by reducing the noise present within the image using Adaptive NLM Algorithm and contrast is enhanced using CLAHE algorithm. The Region of Interest (ROI) to be segmented in contrast-enhanced image is compared with the same in contrast image obtained from the subject injecting the contrast agent. The segmentation performance is compared in terms of its characteristics such as Sensitivity or Recall, Specificity, DICE and Jaccardian Coefficients.

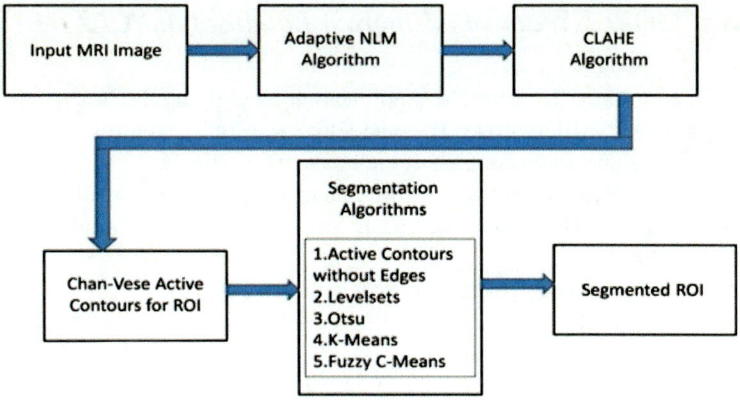

Fig. 9.2 Block diagram of the proposed methodology

Preprocessing of MRI Images

There are several image denoising techniques available, then Non-local filters are widely used in the area of MR imaging [18, 19].

Advantages of NLM Algorithms

- The NLM noise reduction algorithm, in particular, was devised to minimise the loss of underlying image information while selectively removing only the noise.
- After setting areas with the same sized mask positioned around the region of interest, this technique assesses the similarity of the intensity and edge information in a picture (ROI).
- Furthermore, the higher the allocated weight employed during picture processing, the higher the degree of similarity.

In terms of PSNR, MSE and SSIM, experimental results shown in [20] on standard images indicate that the adaptive NLM algorithm outperforms the traditional NLM algorithm.

Contrast Limited Adaptive Histogram Equalisation (CLAHE)

A CLAHE algorithm [21] limit the amplification by clipping the histogram. It has two key parameters: Block Size (BS) and Clip Limit (CL). These two parameters improve the brightness and contrast of the image, respectively.

The CLAHE method to enhance the original image consists of the following steps:

Step 1: initially the input Image is dividing into non-overlapping contextual regions equal to $M \times N$.

Step 2: Histogram is calculated for each contextual region.

Step 3: Calculate the Clip Limit (CL) for each region as given below

$$N_{\text{avg}} = \frac{N_r(x)N_r(y)}{N_{\text{gray}}} \tag{9.1}$$

where N_{avg} is the average number of pixels, N_{gray} is the number of gray levels in the contextual region, $N_r(x)$ and $N_r(y)$ are the numbers of pixels in the X dimension and Y dimension of the contextual region. Then the actual CL is

$$N_{\text{CL}} = N_{\text{clip}} N_{\text{avg}} \tag{9.2}$$

where N_{clip} is the normalised CL in the range of [0, 1]. If the number of pixels is greater than N_{CL} , the pixels will be clipped. The total number of clipped pixels is defined as $N_{\sum \text{clip}}$ then the average of the remain pixels to distribute to each gray level is

$$N_{\text{avggray}} = \frac{N_{\sum \text{clip}}}{N_{\text{gray}}} \tag{9.3}$$

The histogram clipping rule is given by the following statements

$$\text{If } H_{\text{region}}(i) > N_{\text{CL}} \text{ then } H_{\text{region\_ clip}}(i) = N_{\text{CL}} \tag{9.4}$$

$$\text{Else if } H_{\text{region}}(i) + N_{\text{avggray}} > N_{\text{CL}} \text{ then } H_{\text{region\_ clip}}(i) = N_{\text{CL}} \tag{9.5}$$

$$\text{Else } H_{\text{region\_ clip}}(i) = H_{\text{region}}(i) + N_{\text{CL}} \tag{9.6}$$

where $H_{\text{region}}(i)$ and $H_{\text{region\_ clip}}(i)$ are original histogram and clipped histogram of each region at *i*-th gray level.

Step 4: Redistribute the remaining pixels until the remaining pixels have been all distributed. The step of redistribution of pixels is given by

$$\text{Step} = \frac{N_{\text{gray}}}{N_{\text{remaining}}} \tag{9.7}$$

where $N_{\text{remaining}}$ is the remaining number of clipped pixels. *Step* is positive integer at least 1.

Step 5: Then intensity values are distributed according to Rayleigh distribution to enhance the contrast and re-scaled using linear contrast stretching.

Step 6: To eliminate artefacts in the image, new gray level is calculated and assigned to the pixels in contextual region using bilinear interpolation.

Selection and Segmentation of ROI

The Region of Interest to be segmented within the MRI Image is selected manually and Chan-Vese Active Contours [22] was applied to get the object of the image. Manually the Mask will be drawn surrounding the ROI of the object for both CLAHE-enhanced and Contrast-enhanced images. Then efficiency of algorithm is estimated in terms of image characteristics.

Popular Segmentation Algorithms

Here we have taken the threshold-based and clustering algorithms which are useful for Medical Image segmentation applications. They are simple and popular algorithms, such as Otsu Algorithm [23], Active Contours without edges, Level sets [24], K-Means Segmentation [25] and Fuzzy C-Means Segmentation Algorithm [26]. They all are familiar and very simple to understand. Each algorithm has its own advantages and they are application specific.

Performance Metrics

This paper taken spatial overlap-based metrics [27] into account to evaluate the segmentation performance of each algorithm. These are derived from the cardinalities of confusion matrix such as namely the true positives (TP), the false positives (FP), the true negatives (TN) and the false negatives (FN). Then the derives metrics from these cardinalities are mostly used for segmentation evaluation are given by

$$\text{Recall} = \text{Sensitivity} = \frac{\text{TP}}{\text{FN} + \text{TP}} \tag{9.8}$$

This is also called True Positive Rate (TPR). Similarly, the True Negative Rate (TNR) is given by

$$\text{Specificity} = \text{TNR} = \frac{\text{TN}}{\text{FP} + \text{TN}} \tag{9.9}$$

Then

$$\text{DICE Coefficient} = \frac{2 * \text{TP}}{2 * \text{TP} + \text{FP} + \text{FN}} \tag{9.10}$$

This is called Overlap Index mostly used for validating Medical volume Segmentations. For the generalised segmentation with multiple labels, Jaccardian Index is considered.

$$\text{Jaccard Index} = \frac{\text{DICE Coefficient}}{2 - \text{DICE Coefficient}} \tag{9.11}$$

In this analysis, we have considered the above-mentioned four parameters to concise evaluation of segmentation.

Results and Discussions

The datasets were collected from publicly available TCIA database [28] browsing on the preferences as Imaging Modality: MR, Anatomical site: Kidney and Brain and Species: Homo sapiens excluding Phantoms. Totally there are 1043 subjects that were browsed. There are 82 categories of collections of kidney volumes in different projections and 961 categories of collections of Brain structures were obtained from different MRI machines such as GE MEDICAL SYSTEMS, SIEMENS, TOSHIBA_MEC and Philips Medical Systems with different subject IDs. We are mainly concentrate on Axial and coronal projections and selected 32 volumes with sizes 512 × 512 and 256 × 256, thicknesses of 4 mm, 5 mm,6 mm,7 mm, 8 mm and corresponding the spacing of 1.5 mm,2 mm,2.5 mm,3 mm,6 mm and 8 mm, respectively with and without Contrast media. From image volume sequence the slice with maximum tissue area was obtained from different series and then converted those into jpeg format. Each image used in this paper is mentioned with its ID, sequence number and slice number. The sequence/series number differentiates the projection series.

The kidney image shown in Fig. 9.3a is extracted for male subjects at an age of 41 years using GE MEDICAL SYSTEMS, COR KIDNEY1 PRE sequence with slice number 11 out of 20. It was obtained at a magnetic field strength of 1.5 T and Coronal 90 degrees flip angle projection with repetition time 160 s and echo time 1.344 s. The slice is with 256 × 256 size and the thickness of each slice is 5 mm with spacing 6 mm between them. This image is a non-contrasted image shown in Fig. 9.3i, it is preprocessed through Adaptive denoising algorithm to remove unwanted noise shown in Fig. 9.3ii, then contrast is enhanced and gray levels are distributed uniformly using CLAHE algorithm shown in Fig. 9.3iii. A Chan-Vese Active contour Algorithm is used to select the ROI manually and the proposed algorithms are employed for segmentation for both contrast and non contrast images.

Similarly, the kidney image shown in Fig. 9.4a is extracted for male subjects at an age of 45 years using SEIMENS MEDICAL SYSTEMS, COR KIDNEY1 PRE sequence with slice number 30 out of 34.

It was obtained at a magnetic field strength of 1.5 T and Axial 90° flip angle projection with repetition time 120 s and echo time 1.156 s. The slice is with 256 × 256 sizes and the thickness of each slice is 5 mm with spacing 6 mm between them.

Conclusions

The proposed methodology is a new combination of Adaptive NLM Algorithm to reduce the noise in MRI images and CLAHE Algorithm is used to enhance the image contrast before segmenting the MRI Image. The Region of Interest (ROI) to be segmented in contrast-enhanced image is compared with the same in contrast image obtained from the subject injecting the contrast agent. The segmentation performance is compared in terms of its characteristics such as Sensitivity or Recall, Specificity, DICE and Jaccardian Coefficients as shown in Table 9.1 and their graphical evaluation is shown in Figs. 9.5 and 9.6.

The execution time in Fig. 9.6a shows that Otsu, K-Means and Fuzzy C-Means Algorithms are faster for both Non Contrast and Contrast MRI Images compared to the remaining algorithms. The sensitivity is less for Fuzzy C-Means due to soft computing and it is high for Otsu algorithms as shown in Fig. 9.6b. The specificity is more for contrast images in case of Fuzzy C-Means and it is less for Non Contrast image in case of Level set Segmentation. Finally, the DICE Coefficient and Jaccardian Indexes are similar for both Non Contrast and Contrast MRI Images and they are high for segmentation using Otsu Algorithm.

i. Input Image	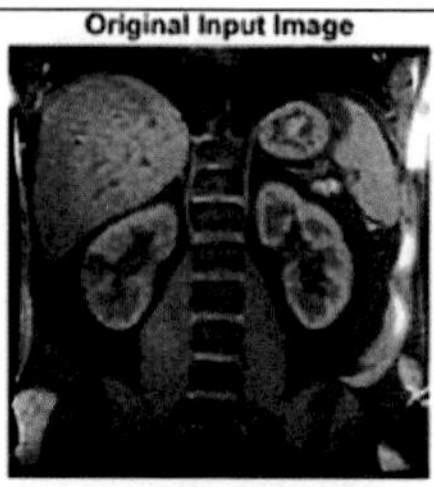 (a) Non Contrast Image	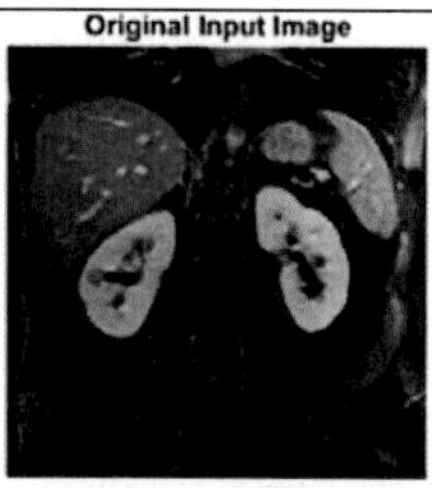 (b) Contrast Image
ii.Denosing Image Using Adaptive NLM Algorithm	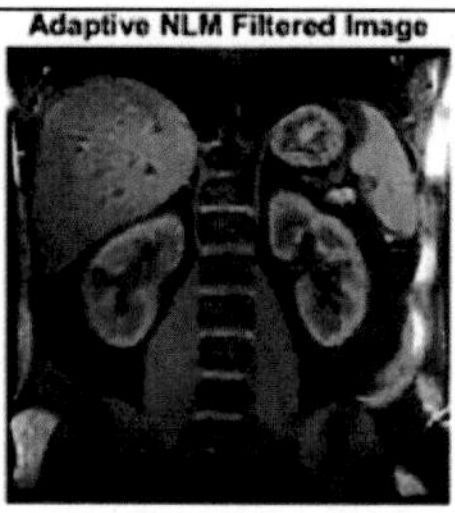 (a) Non Contrast Image	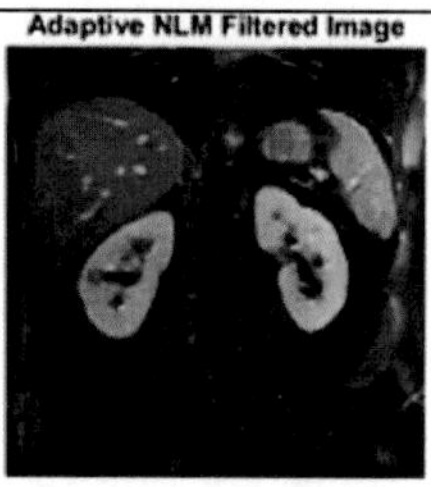 (a) Non Contrast Image
iii.Output of CLAHE Algorithm	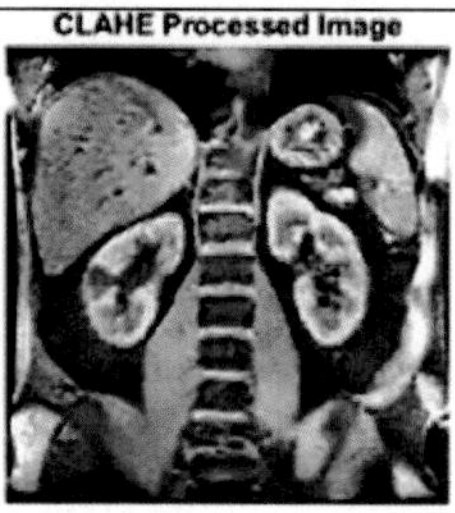 (a) Non Contrast Image	 (a) Non Contrast Image
iv.Mask used for Region Of Interest(ROI)	 (a) Non Contrast Image	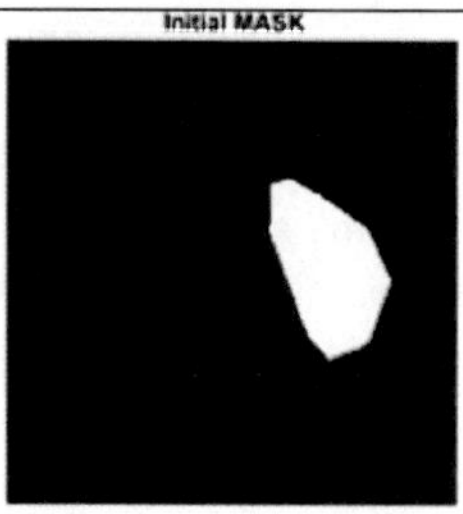 (a) Non Contrast Image

Fig. 9.3 Segmentation process for both non contrast and contrast MRI images-coronal direction

v. Segmentation using Otsu algorithm	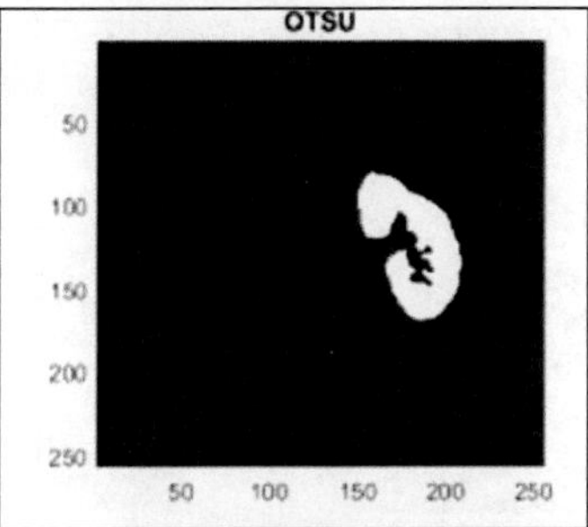	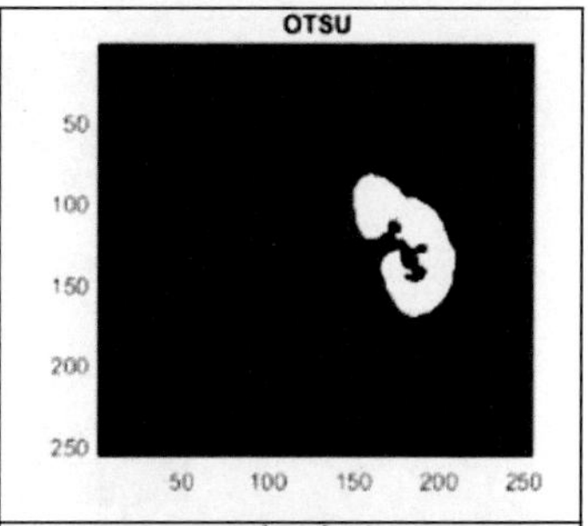
vi. Segmentation using Level sets	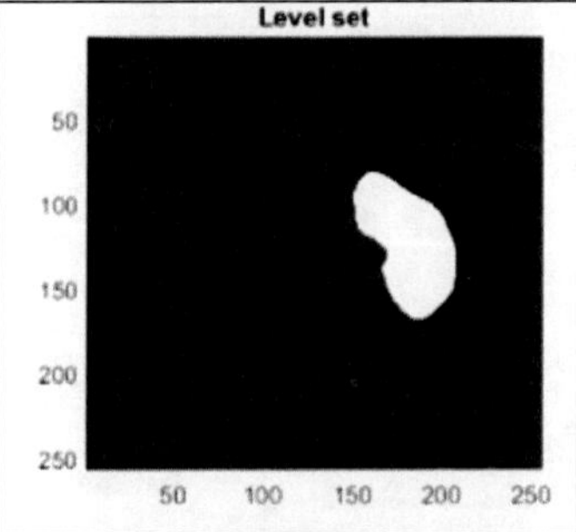	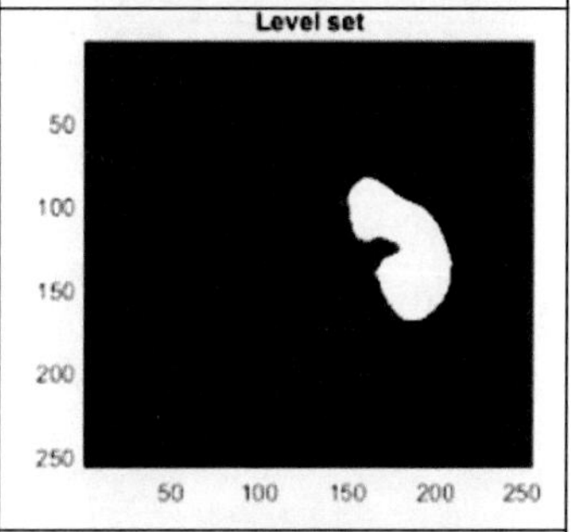
vii. Segmentation using K-Means Algorithm	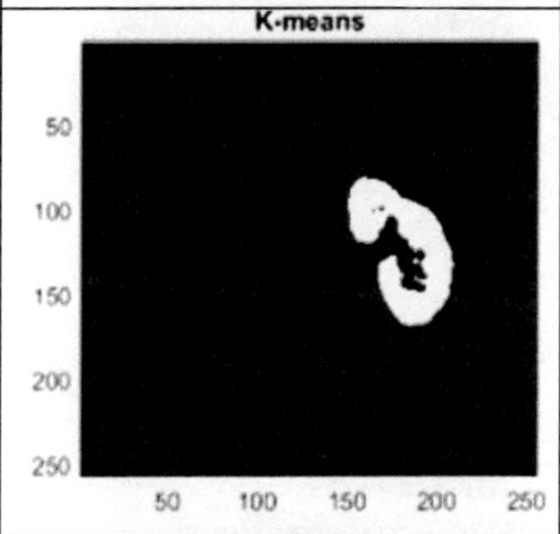	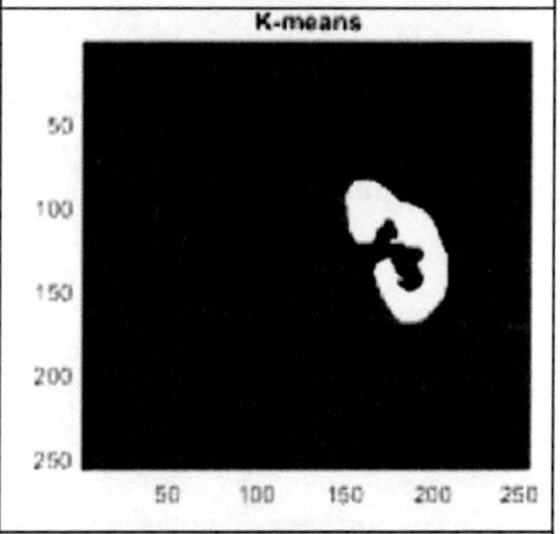
viii. Segmentation using Fuzzy C-Means Algorithm	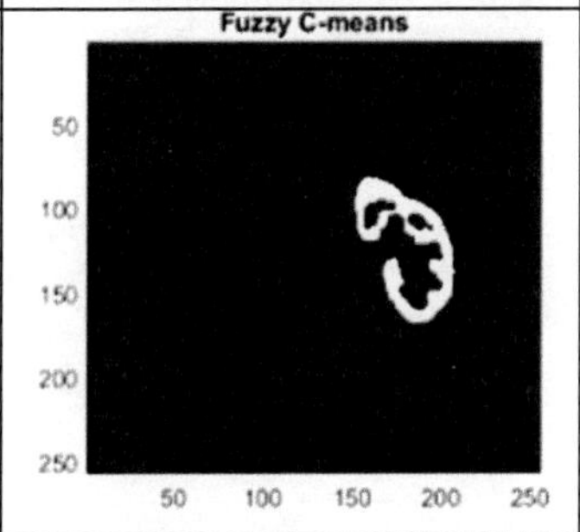	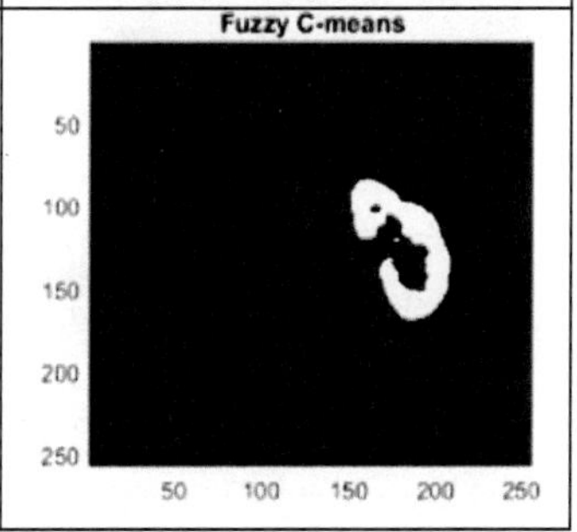

Fig. 9.3 (continued)

i. Input Image	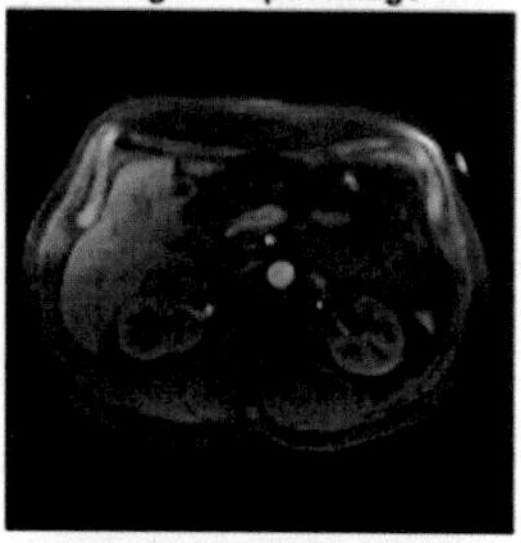	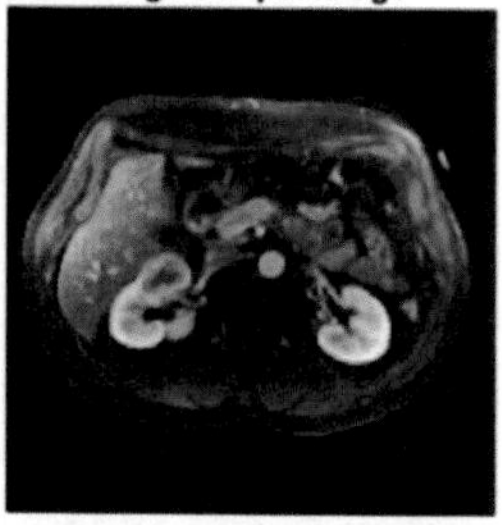
ii. Denoising Image Using Adaptive NLM Algorithm	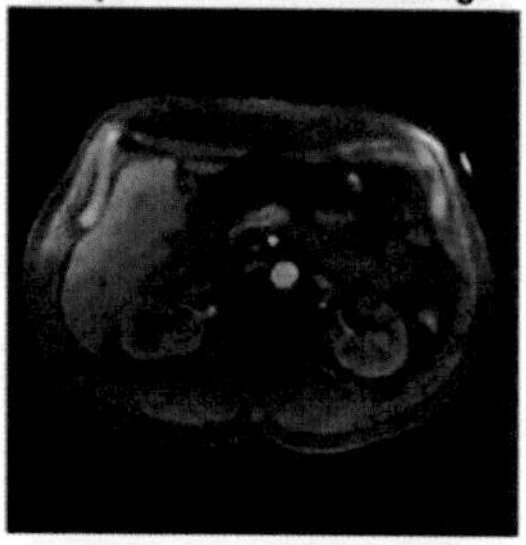	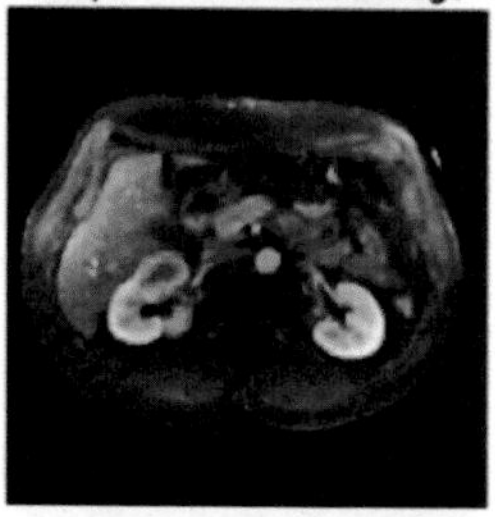
iii. Output of CLAHE Algo-rithm	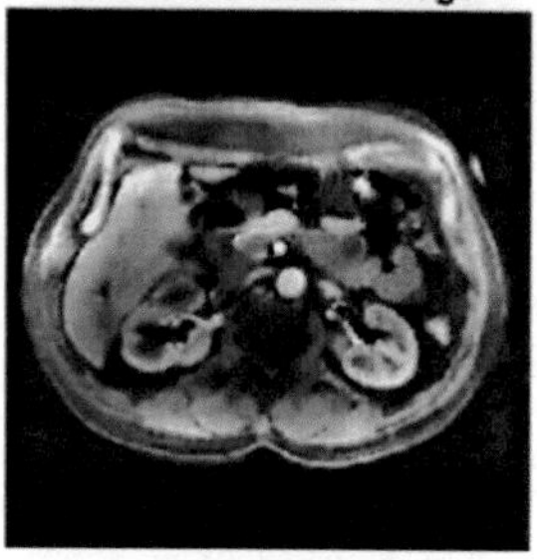	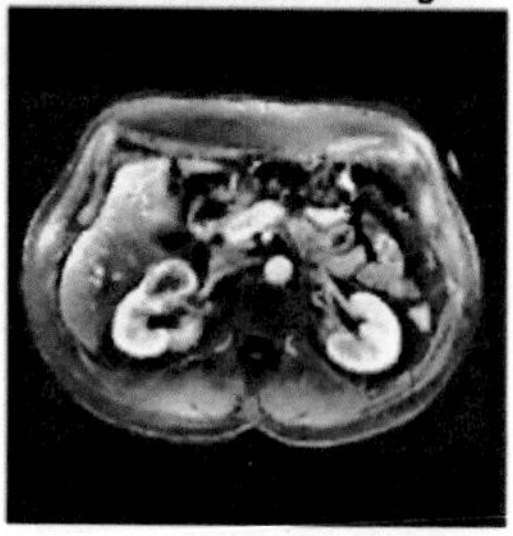
iv. Mask used for Region Of Interest(ROI)		

Fig. 9.4 Segmentation process for both non contrast and contrast MRI images-axial mode

v. Segmentation using Active Contours without Edges	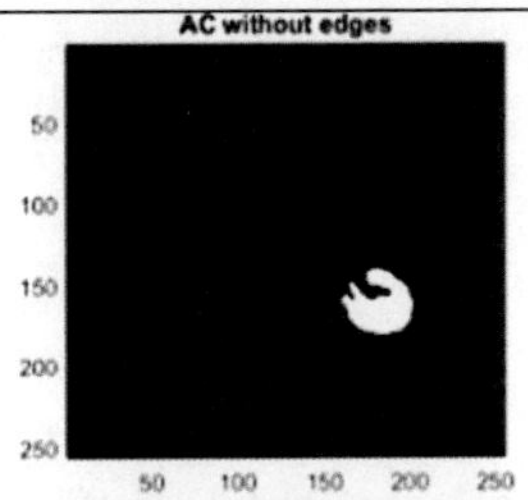	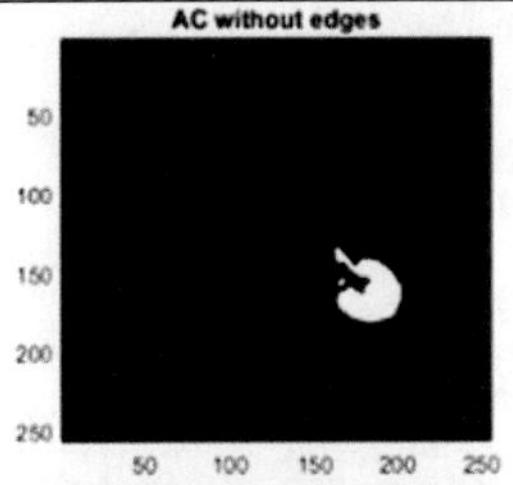
vi. Segmentation using Otsu algorithm	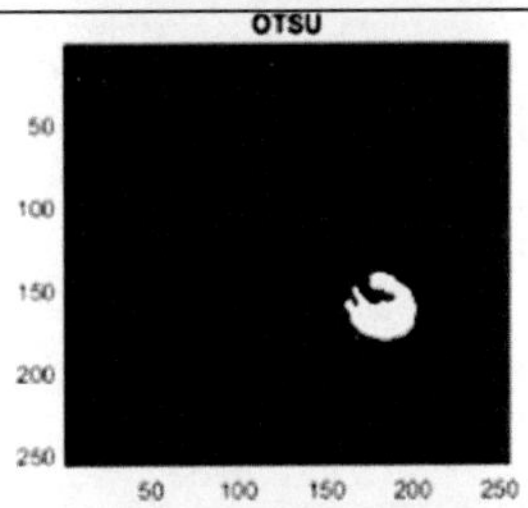	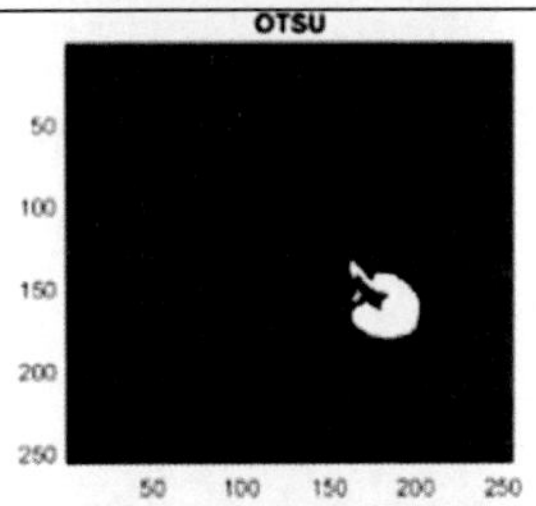
vii. Segmentation using Level sets	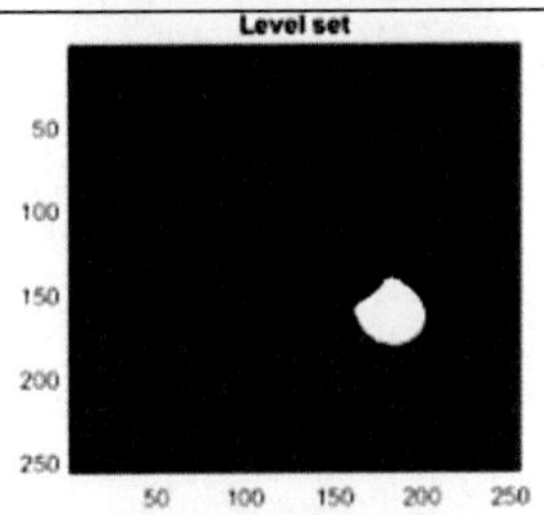	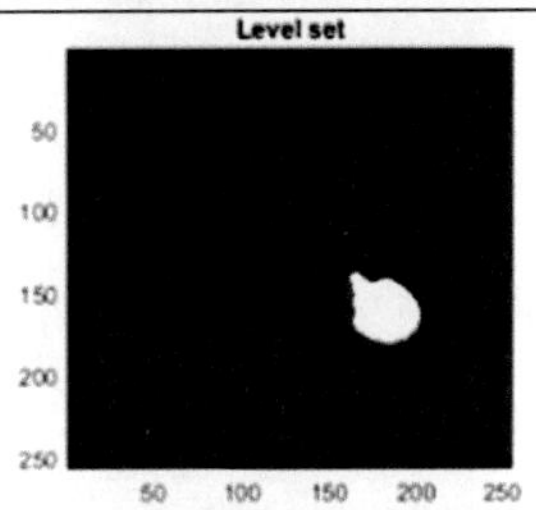
viii. Segmentation using K-Means Algorithm	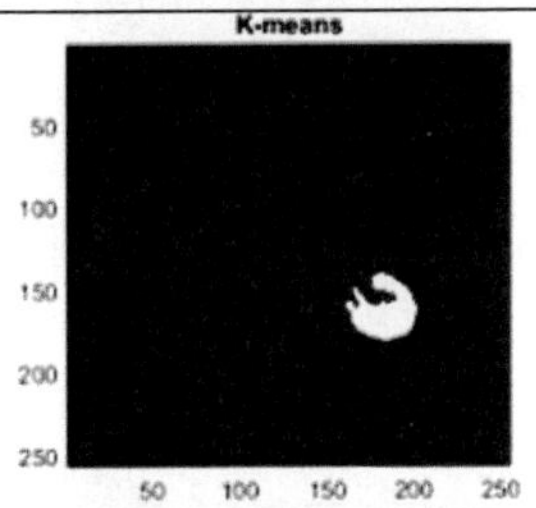	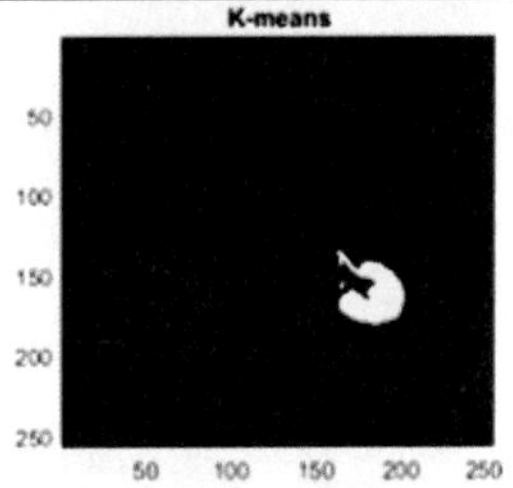
ix. Segmentation using Fuzzy C-Means Algorithm	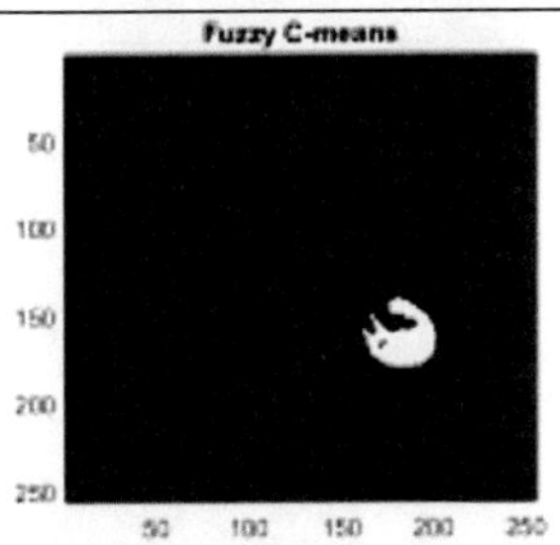	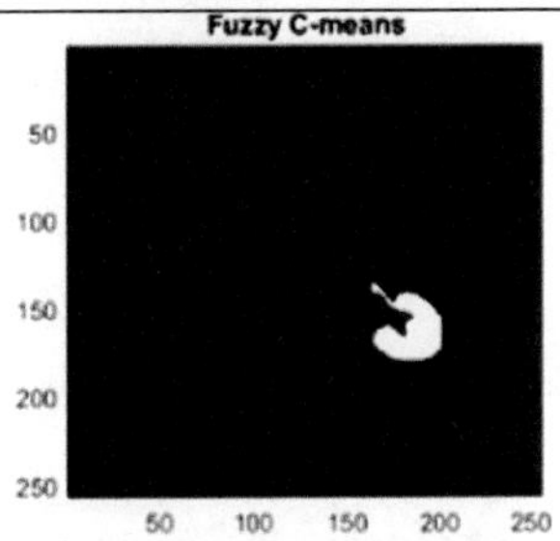

Fig. 9.4 (continued)

Table 9.1 Performance metrics comparison of non contrast and contrast MRI images (for Figs. 9.3 and 9.4)

Input MRI Image	Perf. metric	AC w/o Edges	Level sets	Otsu	K-Means	Fuzzy C-Means
Kidney 4807–11-11-WOC.jpg (Coronal Direction - without Contrast)	Time(in sec)	12.039859	24.186131	0.230866	0.25253	1.960667
	Sensitivity	0.9831	0.9482	0.9841	0.925	0.661
	Specificity	0.9975	0.992	0.9987	1	1
	DICE	0.9635	0.8904	0.9776	0.9605	0.7959
	Jaccard Index	0.9296	0.8024	0.9562	0.924	0.661
Kidney 4807-14-11-WC.jpg (Coronal Direction - with Contrast)	Time(in sec)	9.937055	22.565294	0.047243	0.286961	1.417538
	Sensitivity	0.9828	0.9386	0.9863	0.9016	0.8047
	Specificity	0.9991	0.9946	0.9993	1	1
	DICE	0.9815	0.9114	0.9859	0.9483	0.8918
	Jaccard Index	0.9637	0.8372	0.9722	0.9016	0.8047
Kidney 4989-13-45-WOC.jpg (Axial Direction - without Contrast)	Time(in sec)	9.411271	21.725564	0.02935	0.176253	1.816574
	Sensitivity	0.9885	0.9387	0.9933	0.9788	0.8949
	Specificity	0.9992	0.9976	0.9992	0.9997	1
	DICE	0.969	0.8987	0.9739	0.9807	0.944
	Jaccard Index	0.9399	0.816	0.9491	0.9621	0.894
Kidney 4989-14-45-WC.jpg (Axial Direction - with Contrast)	Time(in sec)	9.554764	21.924253	0.045166	0.199537	1.476591
	Sensitivity	0.976	0.9388	0.984	0.9547	0.8715
	Specificity	0.9996	0.9967	0.9997	1	1
	DICE	0.9769	0.8828	0.9832	0.9755	0.9313
	Jaccard Index	0.9549	0.7901	0.9669	0.9522	0.8715

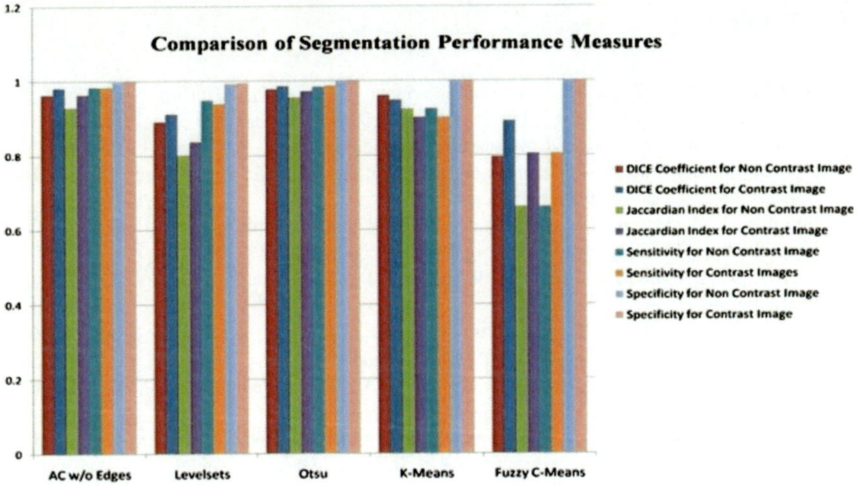

Fig. 9.5 Graphical representation of segmentation performance metrics

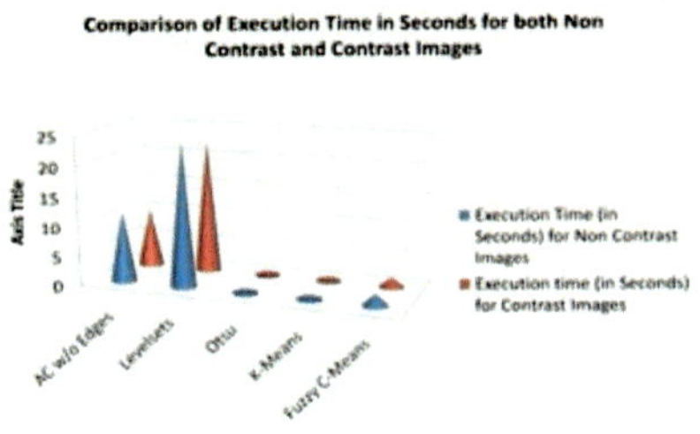

(a) Execution Time in Seconds

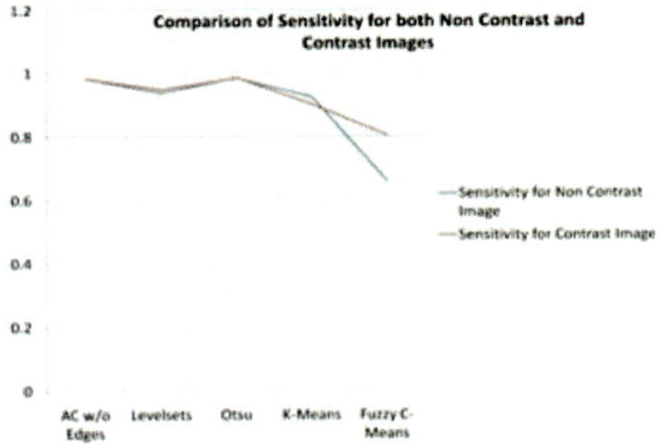

(b) Sensitivity

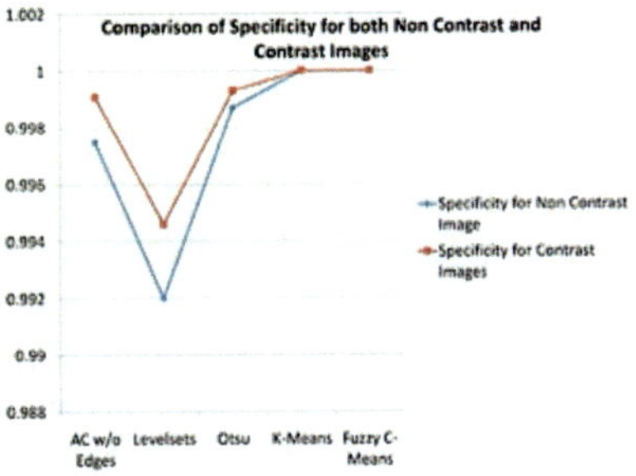

(c) Specificity

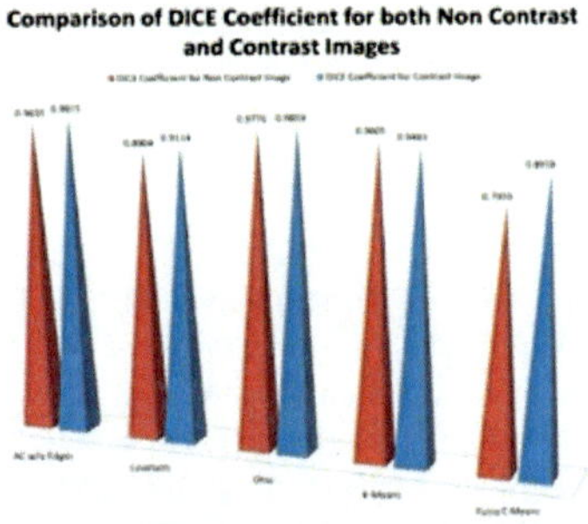

(d) DICE Coefficient

Fig. 9.6 Graphical representation of individual segmentation performance metrics for non contrast and contrast images

References

1. Abdallah, Y.: Improvement of sonographic appearance using HATTOP methods. Int. J. Sci. Res. (IJSR) **4**(2), 2425–2430 (2015). https://doi.org/10.14738/jbemi.55.5283
2. Abdallah, O.M.Y., Alqahtani, T.: Research in Medical Imaging Using Image Processing Techniques, Medical Imaging—Principles and Applications, Yongxia Zhou, Intech Open (2019). https://doi.org/10.5772/intechopen.84360
3. Sharma, N., Aggarwal, L.M.: Automated medical image segmentation techniques. J. Med. Phys. **35**(1), 3–14 (2010). https://doi.org/10.4103/0971-6203.58777
4. Morris, P.: Biomedical Imaging, Woodhead Publishing, pp. xix–xxi (2014). ISBN 9780857091277
5. Hiorns, M.P.: Imaging of the urinary tract: the role of CT and MRI. Pediatr. Nephrol. **26**(1), 59–68 (2011). https://doi.org/10.1007/s00467-010-1645-4
6. Grattan-Smith, J.D., Jones, R.A.: MR urography in children. Pediatr. Radiol. **36**, 1229–1232 (2006). https://doi.org/10.1007/s00247-006-0222-2
7. Weishaupt, D., Koechli, V.D., Marincek, B.: How Does MRI Work?: An Introduction to the Physics and Function of Magnetic Resonance Imaging. Springer Science & Business Media (2013). ISBN 978-3-662-07805-1
8. Ljunggren, S.: A simple graphical representation of fourier-based imaging methods. J. Magn. Reson. **54**(2), 338–343 (1983)
9. Yankeelov, T.E., Gore, J.C.: Dynamic contrast enhanced magnetic resonance imaging in oncology: theory, data acquisition, analysis, and examples. Curr. Med. Imaging Rev. **3**(2), 91–107 (2009). https://doi.org/10.2174/157340507780619179
10. Bailey, D.L., Townsend, D.W., Valk, P.E., Maisy, M.N.: Positron Emission Tomography: Basic Sciences. Springer, Secaucus, NJ. 978-1-85233-798-8 (2005)
11. Azad, G.K., Siddique, M., Taylor, B., Green, A., O'Doherty, J., Gariani, J., et al.: 18F-Fluoride PET/CT SUV? J. Nucl. Med. **60**(3), 322–327 (2019)
12. Galway, K., Black, A., Cantwell, M., Cardwell, C.R., Mills, M., Donnelly, M.: Psychosocial interventions to improve quality of life and emotional wellbeing for recently diagnosed cancer patients. Cochrane Database Syst. Rev. **11**, CD007064 (2012). https://doi.org/10.1002/14651858.cd007064
13. Dementia: Quick Reference Guide (PDF). (UK) National Institute for Health and Clinical Excellence, London. November 2006. ISBN 978-1-84629-312-2
14. Thomas, G., et al.: Measuring the mechanical properties of living cells using atomic force microscopy. J. Visualized Experiments: JoVE **76**, 50497 (2013). https://doi.org/10.3791/50497
15. Jimmy, L.S., Karpiouk, A.B.,, Wang, B., Emelianov, S.Y.: Photoacoustic imaging of clinical metal needles in tissue. J. Biomed. Opt. **15**(2), 021309 (2010). https://doi.org/10.1117/1.3368686
16. Raghavendra, P., Pullaiah, T.: Advances in Cell and Molecular Diagnostics. Academic Press, pp. 85–111 (2018). ISBN 9780128136799. https://doi.org/10.1016/B978-0-12-813679-9.00004-X
17. Meloni, M.F., Smolock, A., Cantisani, V., et al.: Contrast enhanced ultrasound in the evaluation and percutaneous treatment of hepatic and renal tumors. Eur. J. Radiol. **84**(9), 1666–1674 (2015)
18. Bhujle, H.V., Vadavadagi, B.H.: NLM based magnetic resonance image denoising—a review. Biomed. Sign. Process. Control **47**, 252–261 (2019)
19. Buades, A., Coll, B., Morel, J.-M.: A non-local algorithm for image denoising. IEEE Comp. Soc. Conf. Comp. Vis. Pattrn. Recog. **2**, 60–65 (2005)
20. Verma, R., Pandey, R.: Non local means algorithm with adaptive isotropic search window size for image denoising. Ann. IEEE India Conf. (INDICON) **2015**, 1–5 (2015). https://doi.org/10.1109/INDICON.2015.7443193
21. Ma, J., Fan, X., Yang, S.X., Zhang, X., Zhu, X.: Contrast limited adaptive histogram equalization-based fusion in YIQ and HSI color spaces for underwater image enhancement. Int. J. Pattern Recognit. Artif. Intell. **32**, 1–26 (2018)

22. Chan, T.F., Vese, L.A.: Active contours without edges. IEEE Trans. Image Process. **10**(1), 266–277 (2001). https://doi.org/10.1109/83.902291
23. Wang, H., Dong, Y.: An improved image segmentation algorithm based on Otsu method. In: International Symposium on Photoelectronic Detection and Imaging 2007: Related Technologies and Applications, vol. 6625 (2008)
24. Li, C., Huang, R., Ding, Z., Gatenby, J.C., Metaxas, D.N., Gore, J.C.: A level set method for image segmentation in the presence of intensity inhomogeneities with application to MRI. IEEE Trans. Image Process. **20**(7), 2007–2016 (2011). https://doi.org/10.1109/TIP.2011.2146190
25. Venkatachalam, K., Reddy, V.P., Amudhan, M., Raguraman, A., Mohan, E.: An implementation of K-means clustering for efficient image segmentation. In: 2021 10th IEEE international conference on Communication Systems and Network Technologies (CSNT), pp. 224–229 (2021). https://doi.org/10.1109/CSNT51715.2021.9509680
26. Rahman, T., Islam, M.S.: Image segmentation based on fuzzy C means clustering algorithm and morphological reconstruction. In: 2021 International Conference on Information and Communication Technology for Sustainable Development (ICICT4SD), pp. 259–263 (2021). https://doi.org/10.1109/ICICT4SD50815.2021.9396873
27. Taha, A.A., Hanbury, A.: Metrics for evaluating 3D medical image segmentation: analysis, selection, and tool. BMC Med. Imaging **15**, 29 (2015). https://doi.org/10.1186/s12880-015-0068-x
28. TCIA Database. https://nbia.cancerimagingarchive.net/nbia-search/

Chapter 10
A Novel Curve-Energy Framework to Find the Shortest Possible Lines Through Computer Vision

Chandra Sekhar Akula, Asadi Srinivasulu, and Ch. Prathima

Introduction

Several steps are taken to address such flaws by integrating various region-based characteristics to snake-based partitioning algorithms with the motive of achieving them stronger resilient and free of beginning circumstances [1, 2]. Inside one adaptive bounding framework, a method for clinical picture recognition that combines edge and region-based data techniques were discussed [3].

This framework comprises previous knowledge of the shaped form, a limit term that propagates the curve towards sites with high gradient levels, and maybe an area that includes certain area-based data into the boundary-finding structure [4–7].

However, a snake/balloon framework is utilized in this method for classification, and boundary-based data are never utilized because the snake/balloon structure is simply used to place a consistency restriction on the segmentation stage [8].

C. S. Akula (✉)
Avanthi Institute of Engineering and Technology, Vizianagaram District, India
e-mail: chanduforu@gmail.com

A. Srinivasulu
BlueCrest University, 1000 Monrovia, Liberia

Ch. Prathima
School of Computing, Mohan Babu University, Tirupathi, India
e-mail: chilukuriprathi@gmail.com

Ch. Satyanarayana et al. (eds.), *Proceedings of the International Conference on Computer Vision, High Performance Computing, Smart Devices and Networks*, Advanced Technologies and Societal Change,
https://doi.org/10.1007/978-981-19-4044-6_11

Related Works

Furthermore, because the curve dispersion is implemented using a Lagrangian technique, topology alterations are not treated organically [9]. This method, on the other hand, solves the issue by incorporating a region-growing stage. As an outcome, regions with comparable borders are combined if the new region's volatility would be less than the total of the regions' entropies combining.

The "dividing" topological modification would be outside the scope of this step. The work of [10] is limited to binomial distribution and three image classification, whereas it is limited to guided image analysis with a specified number of locations and has been dependent on the transmission for conditionally incompatible curves [11]. Moreover, while data are produced and continually updated over areas, these frameworks are never used any boundary-based data and seem to be extremely sensitive to the beginning circumstances [2]. They are also limited to bi-modal/three-modal or monitored picture classification scenarios.

Proposed Model

The fundamental assumption which is used to develop the proposed framework is that all divisions are equally possible. This condition is never applicable in general, but it has no bearing on the proposed system because the identical framework could be retrieved by substituting the joint distribution of the posteriori frames division likelihood. It is worth noting that maximizing combined segment likelihood seems to be a common optimization criterion. The proposed model is successfully generalized by considering the presence of variables that encapsulate the border and region features of distinct areas of a particular application. The steepest descent technique is used to minimize the generalized optimal solution, resulting in a network of N equation of motion as follows:

$$\begin{cases} \forall x \in [1, N] \\ \frac{\partial v_x}{\partial t} = \alpha(r_x(v_x) - r_{oi}(v_x)N_x(v_x)) + (1-\alpha)(b_x(v_x)K_x(v_x) - \nabla b_x(v_x).N_x(v_x))N_x(v_x) \end{cases} \tag{10.1}$$

The conceptual methodology is used in the following way: The region coordinates are first initialized using a series of random curves. Then, towards the division of the last frame, every region gets distorted according to the associated movement Eq. (10.1). Because for a given pixel that would be ascribed to two locations, pressures having opposite polarity emerge in the associated equation of motion, and the connection between both the regions' locations is derived through the region-based pressure.

The task of image classification is addressed for a synthesized framework made of two categories to show the conceptual approach. The border data are obtained in

Fig. 10.1 Geodesic active regions

this example by applying a Gaussian edge detection algorithm to the gradient value space's norm. Figure 10.1 shows the development of the curves for the category [hA] having regard to three different initiation stages.

The very first [hB] seems to be the background, and it has been made up of pixels whose brightness follows a Gaussian distribution with a value of 150 and even an interquartile range of 10. The 2nd [hA] designates a four region having images with an intensity that follow a Gaussian distribution with an average of 90 and a sample variance of 10.

Humans also present the development of the terms of energy through the period for category A to highlight the proposed new framework freedom from the beginning circumstances. For the very first two initiation phases, the boundary-based power phrase, the region-based power phrase, and also the associated value of the amount of energy are displayed in Fig. 10.2.

Results and Discussion

The employment of template matching techniques provides an extremely elegant technique for propagation curves, in which their location is retrieved by searching

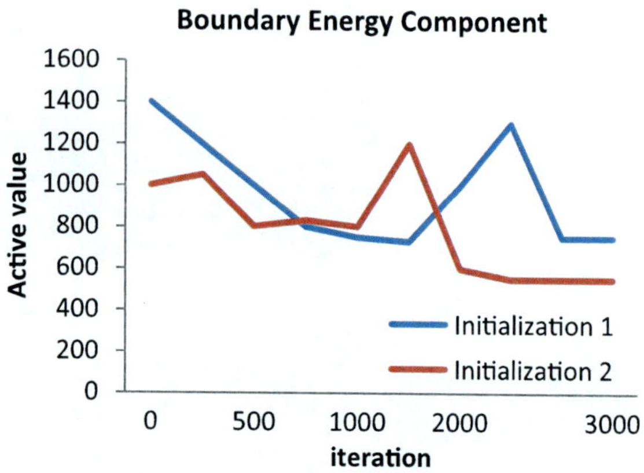

Fig. 10.2 Evolution of the geodesic active region energy

for zero levels setting border crossings. Furthermore, the condition of a specific pixel about an area hypothesis may be simply identified because the associated template matching value lies if it corresponds to the area. It does not correspond to it, and on the other side, the equivalent coefficient is positive. Moreover, because we are using a signed distance measure for the generally more stable implementations, humans could go one stage beyond and estimate the distance of every pixel from every curve. These data are helpful in multi-stage curved dispersion scenarios in which there is no crossover between the various curves. Moreover, at a minimum during the initiation phase, a crossover between the various curves seems to be almost unavoidable. There is chance that an image pixel is not assigned to any theory. Let us pretend that a pixel is originally assigned to two separate zones. Then, introducing an imaginary pressure to the associated stage setting equation of motion that penalizes pixels having numerous labels, a restriction that inhibits a scenario like this is simply implemented. Furthermore, a comparable pressure is used to determine circumstances in which pixels are not assigned to any areas. This is accomplished by altering the equation of motion Eq. (10.2) as follows, as based on the works done [12]

$$\begin{cases} \forall x \in [1, N] \\ \frac{\partial \phi_x}{\partial t}(v) = \beta(r_x(v_x) - r_{oi}(v_x))(\nabla\phi_x(v_x)) + \gamma \sum\limits_{j \in [1,N]} H_x\big(x, \phi_y(s)\big)(\nabla\phi_x(v_x)) \\ +\delta\Big(b_x(v)K_x(v)(\nabla\phi_x(v)) + \nabla b_x(v).\frac{\nabla\phi_x(v)}{(\nabla\phi_x(v))}\Big) \end{cases} \tag{10.2}$$

Humans now apply the very same operations to the input image and obtain an observable collection that is consistent with the texture features. The likelihood of being on the boundary between two different proposed features are calculated for every pixel. Humans need a probability vector as it is dealing with multimodal information. The boundary-based texturing data are displayed by combining the elements of these vectors into a single image using various reliability analysis.

Furthermore, researchers generate region-based information gained from one of the most likely periodic roughest allocations utilizing textural descriptors and also observations collection. The segmentation problem is asserted using the geodesic active region prototype, which aims to find the better duration geodesic curves while maintaining strong bounding likelihoods and generates areas with the highest posteriori categorization likelihood about the accompanying pattern theory. The defined standard value can be calculated using a gradient-descent method, in which the beginning curves are transformed utilizing speeds that cover three conditions: one which gets smaller or broadens the curve towards the corresponding locations, one which endorses the homogeneity of the room curve area provided the affiliated texture theory, and one which wants to express the anticipated curve spatial information (ultra-smooth, sort of consistency). These movement equations are then solved utilizing template matching techniques, with the concept of conditionally incompatible spreading curves imposed by a linked multi-stage propagation. Figure 10.3 depicts some of the experimental outcomes achieved using the proposed framework.

As an outcome, the moving estimate and monitoring problems are connected in this framework, resulting in a two-direction multimodule forward propagating monitoring system that recovers both the movement control and also the actual objective logic at the same time. To conclude, movement prediction and monitoring issues are reformed using the geodesic active region framework, which includes boundary and region-based components. The boundary-based data are calculated using a gradient and perhaps a probability borders detector, whereas the region-based data are calculated utilizing three separate components: movement tracking, intensity-based, and

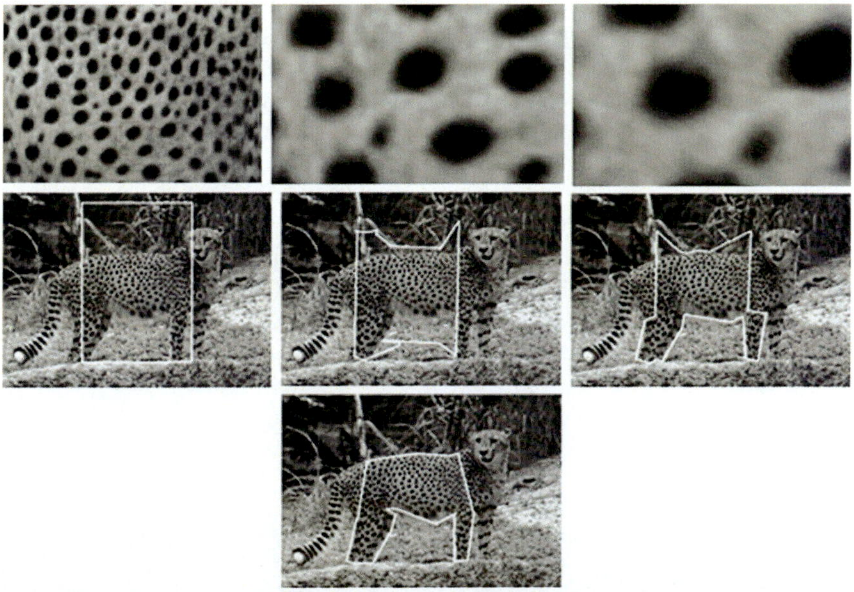

Fig. 10.3 Geodesic active regions for supervised texture segmentation

visual coherence. The variance sampling distribution obtained is analysed as a combination concentration of Gaussian or Laplacian components corresponding to static and movement pixels. The motion detection module is defined using the outcome of this standard statistical step. Humans develop strength monitoring modules in addition to employing certain previous knowledge and changing brightness attributes. Lastly, using the present curved location, a linear movement model can be estimated that provides information about the overall previous and present frames object brightness. To use stochastic gradient descent, the defined objective function derived reference to the curved location and also the movement variables. The derived fractional differential movement formula that modifies the starting curves is restricted by a consistent pressure and is made up of boundary, strength, and motion-based pressures. The template matching methodology is applied to construct those PDEs, and an incremental technique is used to improve the platform-provided estimations. A multi-stage curved propagating is therefore proposed to deal with occlusions, which relates the transmission curves concerning separate elements.

Conclusions

In summary, this research proposes a new vibrational paradigm for handling frame partitioning issues related to energy reduction. The geodesic active region framework is applied on a very flexible paradigm that combines boundary- and region-based data components below a general optimization problem. The curve dissemination theory stems from the enhancement of this structure, in which a set of preliminary curves is perpetuated towards to the geographic locations limits while still being constricted by internal pressures, and it is used to ascertain a coherent separation of the given image based on certain region-based characteristics. Furthermore, researchers had proposed utilizing levels setting means to accomplish curve transmission, which is a very graceful tool that allows for huge number of desirable characteristics while also making curve spreading easier. These strategies, when paired with the proposed model, result in a paradigm that has a wide range of applicability, are free of initial requirements, and also have tremendous power. The introduction of a phrase that allows for some shaped previous information concerning the desired frames division map into the framework and also the expansion of the use of the proposed scheme to deal with applications in three components are difficulties for future developments of this work.

References

1. Yu, H., He, F., Pan, Y.: A novel segmentation model for medical images with intensity inhomogeneity based on adaptive perturbation. Multimedia Tools Appl. **78**(9), 11779–11798 (2019)
2. Vyas, S., Chen, T.J., Mohanty, R.R., Jiang, P., Krishnamurthy, V.R.: Latent embedded graphs for image and shape interpolation. Comput. Aided Des. **140**, 103091 (2021)
3. Tang, Y., Han, X., Li, Y., Ma, L., Tong, R.: Expressive facial style transfer for personalized memes mimic. Vis. Comput. **35**(6), 783–795 (2019)
4. Yu, C., Schumacher, H., Crane, K.: Repulsive curves. ACM Trans. Graph. (TOG) **40**(2), 1–21 (2021)
5. Zhang, Y., Chen, H., Islander, S.L., Gong, J., Xiong, G., Yang, T., Liu, K.: Hybrid trajectory planning for autonomous driving in highly constrained environments. IEEE Access **6**, 32800–32819 (2018)
6. Nguyen, A.T., Lu, S.H., Nguyen, P.T.T.: Validating and forecasting carbon emissions in the framework of the environmental Kuznets curve: the case of Vietnam. Energies **14**(11), 3144 (2021)
7. Pandey, N.N., Muppalaneni, N.B.: Temporal and spatial feature based approaches in drowsiness detection using deep learning technique. J Real-Time Image Proc **18**, 2287–2299 (2021)
8. Shao, H., Kumar, A., Thomas Fletcher, P.: The Riemannian geometry of deep generative models. In: Proceedings of the IEEE Conference on Computer Vision and Pattern Recognition Workshops, pp. 315–323 (2018)
9. Chaudhury, A., Barron, J.L.: Plant species identification from occluded leaf images. IEEE/ACM Trans. Comput. Biol. Bioinf. **17**(3), 1042–1055 (2018)
10. Neagu, O.: The link between economic complexity and carbon emissions in the European Union countries: a model based on the Environmental Kuznets Curve (EKC) approach. Sustainability **11**(17), 4753 (2019)
11. Latchoumi, T.P., Parthiban, L.: Quasi oppositional dragonfly algorithm for load balancing in cloud computing environment. Wireless Personal Commun. 1–18 (2021)
12. Murshed, M.: LPG consumption and environmental Kuznets curve hypothesis in South Asia: a time-series ARDL analysis with multiple structural breaks. Environ. Sci. Pollut. Res. **28**(7), 8337–8372 (2021)

Chapter 11
A Novel Hybrid Tracking Algorithm for Client–Server Connection Using a Machine Learning Technique

P. Rama Santosh Naidu, P. Satheesh, B. Srinivas, and Venkateswarlu Sunkari

Introduction

Large civil engineering activities, sands and dirt mounds, stone quarry, and land pollution for mineral, copper ore, and mineral deposits both have been instances of earth-moving techniques in the extraction and construction sectors [1]. These rely on heavy machinery and seem to be repetitive; modest decreases in cycle lengths, resulting in significant improvement in efficiency, cost reductions, and carbon-reducing emissions [2, 3]. As a result, fast and effective statistics are essential. Finally, the spatiotemporal recordings are useless in the search for the activities that resulted in aberrant phases [2].

If good lines of sight could be picked and Earth-moving equipment is reasonably recognized in that range, machine eyesight technologies have become a feasible solution for monitoring Earth-moving activity [4]. Machine learning is used to control Industrial machinery in several ways [5]. The first method is to create programs that analyze visual data captured by ordinary technological devices like surveillance

P. Rama Santosh Naidu (✉) · P. Satheesh · B. Srinivas
MVGR College of Engineering, Vizianagaram, India
e-mail: prsnaidu@mvgrce.edu.in

P. Satheesh
e-mail: satish@mvgrce.edu.in

B. Srinivas
e-mail: srinio.b@mvgrce.edu.in

V. Sunkari
School of Information Technology and Engineering, Addis Ababa University, Addis Ababa, Ethiopia
e-mail: v.sunkari@aait.edu.et

Ch. Satyanarayana et al. (eds.), *Proceedings of the International Conference on Computer Vision, High Performance Computing, Smart Devices and Networks*, Advanced Technologies and Societal Change,
https://doi.org/10.1007/978-981-19-4044-6_12

cameras. Due to the low price of modern wireless cameras and rising disk drives, it is becoming routine practice to supervise building sites with CCTV systems [6–8].

Computer vision techniques are becoming a viable solution for supervising Earth-moving operations [4] if suitable fields of vision might be chosen and Earth-moving machinery might be adequately identified in that area. Computer science could be applied in a variety of methods to operate industrial equipment [5]. The initial way is developing software that evaluates visual information collected by common technology equipment such as security cameras. Due to the low cost of current wireless sensors and the increasing storage capacity of hard drives, it had become common practice to monitor construction sites in site surveillance. It seems to be a common method with the CCTV method [6–8].

Related Works

SCIT, a new infrastructure architecture proposed in this study, integrates image processing and analysis components, spatiotemporal thinking, and specialist reasoning into a consistent platform. SCIT is used to identify, monitor, and recognize the equipment in packing operations, and also calculate cycle times [9]. The same item identification and movement translation components are used in both systems, but the monitoring engine is diverse: one used a conventional monitoring motor, while someone used an innovative framework to follow the loaded dump truck [10].

Machine vision methods are particularly recent methods for data collection with lot of promise for earth-moving machine monitoring [11, 12]. Item monitoring and item identification, and also efficiency monitoring of personnel, concrete pour, including construction equipment, have now seen advancements in building surveillance applications. Moreover, there seems to be a considerable distance to go into the studies before such a dependable and automated scheme could be used in the construction sector.

Our study aims to bridge the gap among eyesight structures and earth-moving efficiency measurement techniques, allowing it to acknowledge and educated guess mud load-carrying cycles underneath a wider range of visual circumstances, including various perspectives and also the existence of various types of construction machinery. Furthermore, the method only requires a minor human input during the initialization of the camera viewpoint.

Proposed Methods

Researchers believe that the Mean-shift method is effective for monitoring machinery of the hard and physically busy building project, and also that adding the Particle filter and Kalman filter helps improve its efficiency. Mean-shift monitoring seems to be the Kernel-based technique of looking to a local maximum with a dataset's number density and avoids misfits that are far from the maxima. The Mean-shift trackers are supplied with a rich grayscale of the HOG detector response in the second technique. The highly regarded is colored, while the remainder of the answers are adjusted depending on their HOG recognition rating. The Mean-shift approach, on the other hand, has a drawback: because this algorithm looks of local maxima, a monitoring blob might otherwise move to the distant item with comparable color or texture. This issue, and other researchers found, is discussed in greater detail in the discussion part. The HOG method used in our hybrid method is to track designated devices. Since equipment characteristics do not dramatically change across time steps, this would be possible. When the sensor recognizes that a truck has been loaded, it keeps looking for vehicles, but more effectively with considerably shorter periods. The memory capabilities search for only 3 positions each 2 s within those circumstances, namely the targeted dump truck's initial orientation and two nearby perspectives, which take roughly 0.39 s with the same GPU and processors. The structure, for instance, only searches for rear-left, front-left and side-left perspectives to a targeted truck are oriented leftside. Modifications in a vehicle's trajectory are detected rapidly in this manner, but the computing work required to verify all 8 orientations is avoided by using prior information. Moreover, the threshold is reduced to prevent false negatives, but the frequency of false alarms rises. Every identification builds a bounding box around the possible target position. Figures 11.1 and 11.2 shows the chart and graphical sequencing of the actual process, accordingly.

The SCIT system is built utilizing 2 different algorithms, as previously stated, and this shares information on the architecture of both SCIT systems. They are created in Visual C++ expressive to use the open-sourced OpenCV 2.3.1 package. It transmits the observed boundary to the average monitoring algorithm once it has been located (see Fig. 11.3). SCIT's current design ends scanning after one excavator, but the architecture may be modified to analyze films with 2 or more capacity utilization. The device begins scanning for dump trucks at specific time intervals in addition to implementing the excavators. Because scanning a 640 × 480 pixel image for all 8 directions requires around 1.07 s, the method could be configured for a look to dump trucks at some period longer than 1.07 s. The proposed four gaps are used in this circumstance. Within every identification period, the activity recognition studies reveal all of the observed dump trucks to see if any of them satisfy the categorization. When the software approves the loaded operation, this would cease looking for dump trucks and send the loaded vehicle to the resources and capabilities, which would establish the load area and document the loaded cycle's beginning (this graphic illustrates the Mean-shift monitor). The loading area is specified as 1.25 times the length of the vehicle and 1.5 times the height of the vehicle.

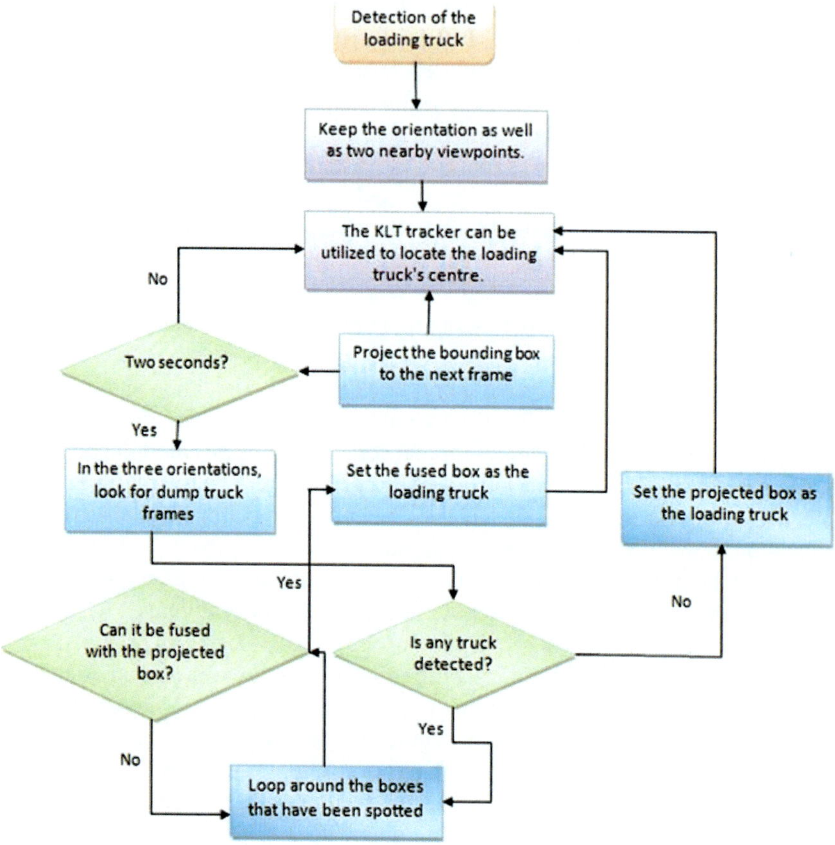

Fig. 11.1 Hybrid tracking method

Fig. 11.2 Finding of truck in the image

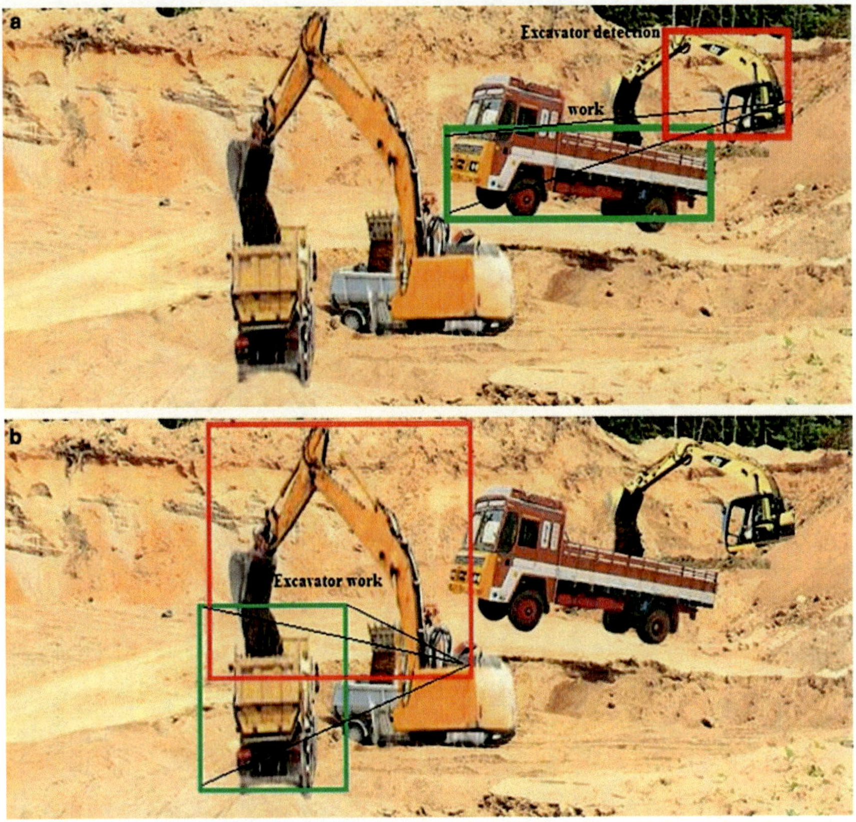

Fig. 11.3 **a** The excavator's detection; **b** the excavator's tracking

Experimental Results

Eighteen (18) recordings with an overall length of 2 h and twenty-seven mins are collected of excavation and construction of two (2) condominium buildings in downtown Toronto, Ontario, to evaluate and compare these two devised methodologies. In the training phase, neither of these movies are used. The criterion for assessing the machinery used in these two tasks is extremely similar, resulting in a homogeneous production database. The movies featured two kinds of hydraulic excavators and also many brands of urban garbage trucks having equal carrying capacities, including Mack, Sterling, Volvo, and Kenworth. These movies are captured by several researchers over 8 site inspections throughout all 4 seasons, underneath different illumination circumstances, from both levels and raised views, using two separate brands of digital photography (see Fig. 11.4).

Fig. 11.4 Views of the gravel process

The excavator is painted in conventional building colors, while the garbage trucks are painted in a selection of colors. SCIT computers used separate action identification criteria to analyze the testing movies. The data of positive instances, false alarm rates, and average processing periods of real positive phases are shown in Table 11.1. The loading period begins whenever a dump truck entirely stops at the front of the excavators, and it concludes whenever the truck begins to move out of the area, according to the human inspection. The human activity recognition threshold is smaller in testing with a lower Identification number, and also the criterion grows as the series of experiments rises.

SCIT with hybrids, testing, 3 had the best result, as indicated in Table 11.1. Test 3's threshold is ideal, it has the most real positive cycles and has lower variance than test 2. The Mean-shift method is used in only one testing session, utilizing the objective criterion established in the Hybrid 3 experiment, but the results are poor, therefore the other trials are canceled.

Table 11.1 Experimental results of sample

Data collection method	No. of detected cycles	False negative cycles	False positive cycles	Average cycle time (based on true positive cycles only) (s)
Manual	56	0	0	101.88
Test 1: SCIT with hybrid	54	0	5	107.50
Test 2: SCIT with hybrid	55	0	3	107.23
Test 3: SCIT with hybrid	55	0	3	105.95
Test 4: SCIT with hybrid	52	4	2	105.4
Test 3a: SCIT with mean-shift utilizing HOG	49	0	3	104.41
Test 3b: SCIT with mean-shift utilizing color features	32	0		105.05

Reducing the requirements raises the risk of false positive, like identifying a false alert item or mistaking a neighboring truck for one which has been loaded. A smaller threshold might even result in long-term cycle times since researchers notice the loaded truck before it's even fully positioned itself for loading. Raising the threshold increased the average amount of time, according to the findings. Furthermore, it enhances productivity up to a point (see Table 11.1) before causing some missed cycles due to the SVM classifier's failure to accept the real loaded vehicle. Figure 11.5 shows the same working location with the same machinery arrangement shot from a greater viewpoint only a few moments later. The SCIT with hybrid trackers accurately monitors the loading vehicle in this situation. The relevance of the camera's right perspective for accurate outcome is demonstrated by these findings. The cameras could be mounted on neighboring structures, overhead cranes, sloping tops, or permanent posts; however, many construction projects do not have this luxury because there are no surrounding overseeing spots or they are inaccessible.

Conclusion

In this paper, researchers provide SCIT, a vision-based system for recognizing and estimating dirt load phases. The prototype program is made using two separate monitoring approaches and would then be tested on multiple test videos taken at two (2) distinct building sites with different lighting conditions. The SCIT for the new

Fig. 11.5 **a** Incorrect detection and tracking **b** correct identification and monitoring

hybrid evaluating algorithm surpassed the traditional Mean-shift tracker. The study introduces a new detection technique and also an activity recognition module for estimating muddied loaded cycles in practice. The technology would be expanded to recognize and analyze various sorts of earth-moving activities with several units in the coming days.

References

1. Hitimana, E., Bajpai, G., Musabe, R., Sibomana, L., Kayalvizhi, J.: Implementation of IoT framework with data analysis using deep learning methods for occupancy prediction in a building. Future Internet **13**(3), 67 (2021)
2. Traini, S., Sciullo, L., Trotta, A., Di Felice, M.: Practical indoor localization via smartphone sensor data fusion techniques: a performance study. In: 2019 16th IEEE Annual Consumer Communications and Networking Conference (CCNC), pp. 1–7. IEEE (2019)

3. Salih, A.A., Ameen, S.Y., Zeebaree, S.R., Sadeeq, M.A., Kak, S.F., Omar, N., et al.: Deep learning approaches for intrusion detection. Asian J. Res. Comput. Sci. 50–64 (2021)
4. KhoKhar, F.A., Shah, J.H., Khan, M.A., Sharif, M., Tariq, U., Kadry, S.: A review on federated learning towards image processing. Comput. Electr. Eng. **99**, 107818 (2022)
5. Yang, S., Hu, J., Jiang, K., Xiao, H., Wang, M.: Hybrid-360: an adaptive bitrate algorithm for tile-based 360 video streaming. Trans. Emerg. Telecommun. Technol. e4430 (2021)
6. Uribe, S., Belmonte, A., Moreno, F., Llorente, Á., López, J.P., Álvarez, F.: New access services in HbbTV based on a deep learning approach for media content analysis. AI EDAM **33**(4), 399–415 (2019)
7. Al-Mashhadi, S., Anbar, M., Hasbullah, I., Alamiedy, T.A.: Hybrid rule-based botnet detection approach using machine learning for analyzing DNS traffic. Peer J. Comput. Sci. **7**, e640 (2021)
8. Wei, X., Zhou, M., Kwong, S., Yuan, H., Wang, S., Zhu, G., Cao, J.: Reinforcement learning-based QoE-oriented dynamic adaptive streaming framework. Inf. Sci. **569**, 786–803 (2021)
9. Latchoumi, T.P., Parthiban, L.: Quasi oppositional dragonfly algorithm for load balancing in cloud computing environment. Wirel. Pers. Commun. 1–18 (2021)
10. Butt, U.A., Mehmood, M., Shah, S.B.H., Amin, R., Shaukat, M.W., Raza, S.M., et al.: A review of machine learning algorithms for cloud computing security. Electronics **9**(9), 1379 (2020)
11. Abbasloo, S., Yen, C.Y., Chao, H.J.: Classic meets modern: a pragmatic learning-based congestion control for the internet. In: Proceedings of the Annual Conference of the ACM Special Interest Group on Data Communication on the Applications, Technologies, Architectures, and Protocols for Computer Communication, pp. 632–647 (2020)
12. Pandey, N.N., Muppalaneni, N.B.: A novel algorithmic approach of open eye analysis for drowsiness detection. Int. J. Inf. Technol. **13**, 2199–2208 (2021)

Chapter 12
Surface Damage Indicators Identification Through Simulation Studies in Computer Vision

B. Srinivas, P. Satheesh, and P. Rama Santosh Naidu

Introduction

The construction of a mathematical formula to define input relationships of a component by evaluating measured data becomes a very basic definition of advanced authentication. The notion of St-Id in engineering structures is investigated in the literature since 1970s [1, 2]. A St-Id method incorporates several characteristics, referred to as a structural damage indicator, that is formed by creating an input–output link that is used to assess changes or damages throughout time to make decisions [2].

A damages indication must give proof when there is a deviation from a structure's predefined or specified healthy quality of life to attain the goal. The modal parameters produced from the descriptive study are conducted and proper parameterization methods [3] have been the most prominent set of harm signals.

Related Works

Natural frequency, mode forms, curvature forms, modality adaptability, dampening ratio, and others are instances of damaged signals in this category, which the researchers applied to real-world constructions [4]. One of the major drawbacks of

B. Srinivas (✉) · P. Satheesh · P. Rama Santosh Naidu
MVGR College of Engineering, Vizianagaram, India
e-mail: srinio.b@mvgrce.edu.in

P. Satheesh
e-mail: satish@mvgrce.edu.in

P. Rama Santosh Naidu
e-mail: prsnaidu@mvgrce.edu.in

Ch. Satyanarayana et al. (eds.), *Proceedings of the International Conference on Computer Vision, High Performance Computing, Smart Devices and Networks*, Advanced Technologies and Societal Change,
https://doi.org/10.1007/978-981-19-4044-6_13

employing modal analysis, according to the researchers, seems to be the requirement for a thick array of accelerometers to produce more accurate damage signals, particularly for recognition and localization [5]. This condition makes ordinary surveillance investigations for real-world structures inconvenient and prohibitively expensive.

On lengthy crossings, crawling examinations or train weight are favored as hazard indications, as shown in various articles [6–8]. For bridges surveillance, a hybrid sensor-camera technique is proposed for calculating strain UILs without the use of a dedicated testing vehicle. Although the use of UILs as damages indications is prospective in BHM, the installation of a single vehicle that would maintain the same path for every information gathering instance limits UIL development [9].

This restricts the use of UIL for BHM in a few specific situations. Eyesight methods are studied as a promising strategy in the condition monitoring sector, thanks to new imaging technology and machine learning methods. Vision-based techniques, without a doubt, enable for non-contact measurement implementation, which enhances surveillance performance and decreases the lump-sum price of structure evaluations [10, 11].

Despite numerous difficulties in vision-based SHM, investigators are using image analysis to solve some of the problems which have been involved with the SHM society, like measuring cracks and delamination on functional substrates, and estimating functional deformations through eyesight methodologies.

Proposed Method

The goal of this research is to propose and establish new architecture standards, and detect damaged indications that have been dependent on vision-based measures. The proposed damages indication is a three-dimensional creation of a component impact collection called a units impact surface. A UIS, like a UIL, is utilized as a probable BHM harm indication. Scientists have demonstrated that employing a UIL had restrictions, such as maintaining only one vehicle traveling along such a pathway. Because it might manage several cars moving over randomized paths, creating a UIS doesn't quite necessitate this kind of implementation. It also allows for the use of transport vehicles for BHM without any need for a crossing shutdown. Various vehicle-loaded data, such as vehicle loads and locations, are approximated in the study using a variety of technological recognition methods. The observed deformations are combined with the input information [12]. The author discusses on gathering, transferring, and utilizing machine learning for non-contact, goal assessments (Fig. 12.1).

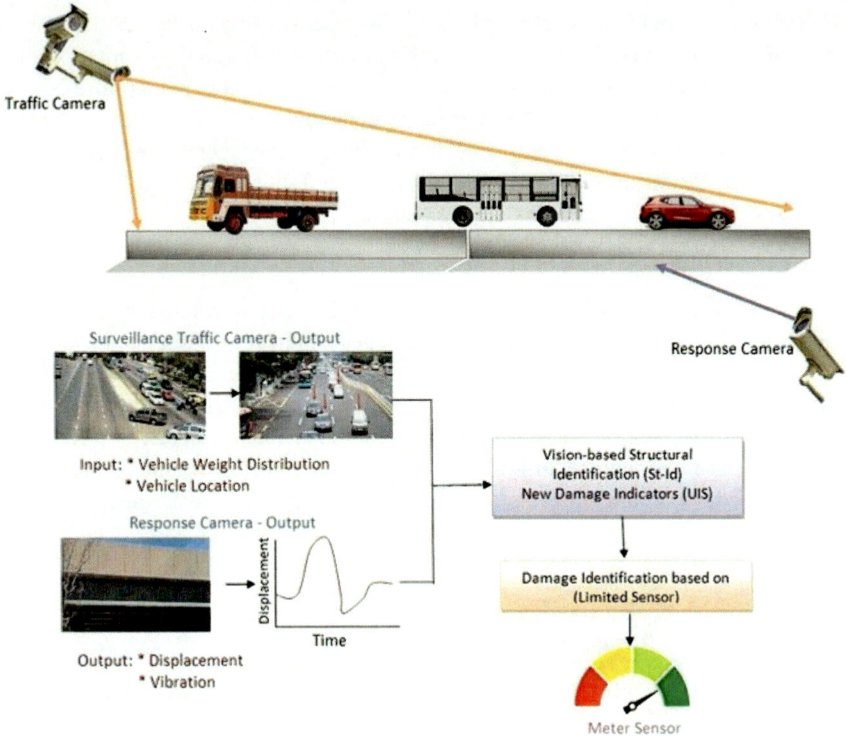

Fig. 12.1 Schematic diagram for the vision-based bridge

Construction of UIS

Measurements that include vehicle axle loads, vehicle positions, and structure deformations are used to create the movement UIS. In a nutshell, the process for building UIS is based on ground techniques for a collection of 3D separate points known as UIS points. The condition of the vehicle loading-structural reaction relationship is described by a UIS point. Innately, this certain condition corresponds to a moment when a security video captures an image of many vehicles and the resulting bridges girder deformation induced by those vehicles [13]. The horizontal value of a UIS point seems to be the normalized movement of the bridge girder, whereas two rectangular variables of a UIS location are tower dimensions of the arrangement site obtained from vehicle placements.

AdaBoost-based cascades classifiers are trained with tag images acquired from positive and negative images to produce a detector. A vehicle detector, for example, is created by training a group of vehicle photographs representing a set of non-truck images, like landscapes or other sorts of vehicles. To produce a detector for every type of vehicle, a commercial AdaBoost-based cascades classification implemented in MATLAB is employed. Figure 12.2 depicts the training program. After obtaining the detectors, it would be developed to determine items using a scanner window in any section of an image. The scanner windows crop the image representation of a specific area, resulting in an image window. The image window is therefore passed into the AdaBoost-based cascades classifier's detectors, which determines whether the elements of the windows are objects or non-objects. The technique is performed for every detector one at a period when there are several sensors for various categories of items. Figure 12.3 depicts the detection strategy.

Vehicle Detection and Classification

To determine the appropriate vehicle categories, every detector created in the learning program is utilized to scan through a source image at multiple graphics and positions. Figure 12.4 shows the genuine favorable results of object recognition. Bounding boxes that correspond to Classes 1–3 are used to categorize the detected cars. Even though most cars may be correctly identified, some erroneous detections could be discovered, as illustrated in Fig. 12.4. Undetectable cars, surrounding areas, the incorrect kind of classification, and erroneous bounding box measurements all seem

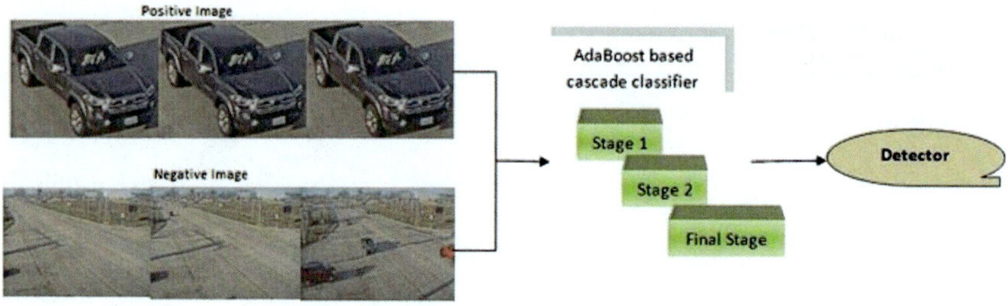

Fig. 12.2 The training process for obtaining a detector

Fig. 12.3 Detection scheme using a trained detector

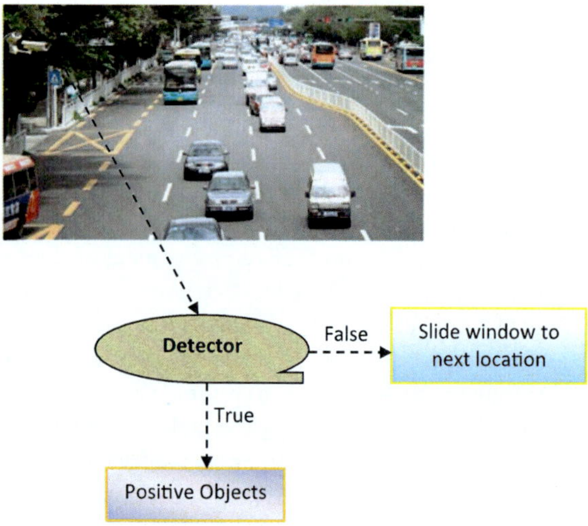

to be examples of false detection. The amount of false detections out of 5826 photos obtained from 40 tests are used to assess the recognition procedure's effectiveness. The quality of all 5826 photographs is physically confirmed since there is no ground-truth image collection of confirmation, the investigators. The assessment is assessed by comparing detection performance between both the 3 classifications (see Table 12.1) and between standard and various options (see Table 12.2). Table 12.1 reveals that the Class 1 detection performance (95.8%) is the best, while the Class 2 detection performance (89.1%) would be the weakest. This is attributed to the fact that the front end of Class 1 cars seems to have more surfaces, making differentiation for the classifiers simpler. During the study, it is discovered that the majority of false detections happen when the vehicles are moved far from the camera. In photos, automobiles that are chased away from the camera frequently appear tiny and hazy things.

The displaced UISs are retrieved from the variety for points on a UCF 4-span bridge girders to compare them and to locate damage. The non-target species vision-based technique are used to determine the movement of a point of measurement. The deflections from measuring places are generated via a set of LVDTs to show UIS for different sites. Since there is only one video system at the present moment, the eyesight, movement provisional specification is used only at the L1 site. Both the nationwide instrumentation, measurement method, and also the webcams for collecting compact cars traveling so over bridge deck are activated from a single

Fig. 12.4 Accuracy

Table 12.1 Detection rates vehicles class

Vehicle class	Detection rates (%)
Class 1	95.8
Class 2	89.1
Class 3	94.6

Table 12.2 Accuracy prediction

No. of vehicles	Detection rates (%)
Single	92.7
Multiple	87.5

LABVIEW program to synchronize the visual information of vehicles and the movement information of the framed structure.

Conclusions

The load and reaction connection is developed by an image-based structure recognition method which provides a unique type of harm indication incorporating structural characteristics of bridge monitoring systems. UIS seems to be the proposed damage indication, which has been created utilizing eyesight measurement results and an incremental approximation technique. In the UCF Structural Lab, the proposed image-based St-Id technology is tested and evaluated on a four-span bridge. The constancy and dependability of the damage indication, UIS, is similar platforms, with the highest CV number of 2.31% and perhaps other CV value mostly less than 1.00% based on a measurement of information dispersal for the UIS. The UIS health indication is utilized to detect and localize harm on a UCF 4-span bridge in a modeled damage situation.

References

1. Liu, Y., Guo, L., Gao, H., You, Z., Ye, Y., Zhang, B.: Machine vision-based condition monitoring and fault diagnosis of machine tools using information from machined surface texture: A review. Mech. Syst. Signal Process. **164**, 108068 (2022)
2. Tu, K., Wen, S., Cheng, Y., Zhang, T., Pan, T., Wang, J., Sun, Q., et al.: A non-destructive and highly efficient model for detecting the genuineness of maize variety 'JINGKE 968' using machine vision combined with deep learning. Comput. Electron. Agric. **182**, 106002 (2021)
3. Dong, C.Z., Catbas, F.N.: A review of computer vision-based structural health monitoring at local and global levels. Struct. Health Monit. **20**(2), 692–743 (2021)
4. Hoskere, V., Narazaki, Y., Spencer, B.F.: Physics-based graphics models in 3D synthetic environments as autonomous vision-based inspection testbeds. Sensors **22**(2), 532 (2022)
5. Xing, L., Wang, Y., Luo, R., Zou, L.: Application of computer vision technology in agricultural products and food inspection. J. Phys. Conf. Ser. **1915**(3), 032045, May 2021 (IOP Publishing)
6. Hashmi, A.W., Mali, H.S., Meena, A., Khilji, I.A., Hashmi, M.F.: Machine vision for the measurement of machining parameters: A review. Mater. Today Proc. (2021)
7. Chen, Q., Lin, H., Zhao, J.: Computer vision technology in food. In: Advanced Nondestructive Detection Technologies in Food, pp 91–126. Springer, Singapore (2021)
8. Yang, A.M., Zhi, J.M., Yang, K., Wang, J.H., Xue, T.: Computer vision technology based on sensor data and hybrid deep learning for security detection of blast furnace bearing. IEEE Sens. J. **21**(22), 24982–24992 (2021)
9. Latchoumi, T.P., Parthiban, L.: Quasi oppositional dragonfly algorithm for load balancing in cloud computing environment. Wirel. Pers. Commun., 1–18 (2021)
10. Qiao, W., Ma, B., Liu, Q., Wu, X., Li, G.: Computer vision-based bridge damage detection using deep convolutional networks with expectation maximum attention module. Sensors **21**(3), 824 (2021)

11. Zdziebko, P., Holak, K.: Synthetic image generation using the finite element method and blender graphics program for modeling of vision-based measurement systems. Sensors **21**(18), 6046 (2021)
12. Huang, M.Q., Ninić, J., Zhang, Q.B.: BIM, machine learning and computer vision techniques in underground construction: Current status and future perspectives. Tunn. Undergr. Space Technol. **108**, 103677 (2021)
13. Pandey, N.N., Muppalaneni, N.B.: Real-time drowsiness identification based on eye state analysis. In: International Conference on Artificial Intelligence and Smart Systems (ICAIS), pp. 1182–1187 (2021). https://doi.org/10.1109/ICAIS50930.2021.9395975

Chapter 13
Gaussian Scale Concept to Reduce the Computation in Detection of Surface Defects in Machine Vision

P. Satheesh, B. Srinivas, and P. Rama Santosh Naidu

Introduction

Visual examination using artificial intelligence and machine learning approaches is increasingly popular over the years for assessing surface flows of manufactured products and materials [1]. Techniques of visual examination are shown to deliver quick statistical evaluations with enhanced performance and productivity at a cheap cost [2].

Turbines are enormous machining cylinders used to fire shells and are mounted massive howitzers. As the bullet explodes and glides into the turbines, temperature and pressure build-up, gradually degrading the surface [3]. The deteriorated surface imperfections may cause the projectile to launch incorrectly, or possibly endanger the lives of those controlling it. As a result, assessing flaws in damaged turbines is critical [4].

On worn turbines, the main kinds of flaws recognized are ordinary wearing, corrosive pits, erosion, and rusting. The development of wearing is due the relative velocity of the projectile and also gun barrel. Destructive pitting and degradation is caused by the impact of combustion [5]. The oxidization of the surface causes rusting. If the damaged region exceeds a certain threshold, the turbines would be repaired or discarded. As a result, the flaws must be discovered, categorized, and also by the region assessed [6]. In this case, mechanical approaches are challenging because

P. Satheesh (✉) · B. Srinivas · P. Rama Santosh Naidu
MVGR College of Engineering, Vizianagaram, India
e-mail: satish@mvgrce.edu.in

B. Srinivas
e-mail: srinio.b@mvgrce.edu.in

P. Rama Santosh Naidu
e-mail: prsnaidu@mvgrce.edu.in

Ch. Satyanarayana et al. (eds.), *Proceedings of the International Conference on Computer Vision, High Performance Computing, Smart Devices and Networks*, Advanced Technologies and Societal Change,
https://doi.org/10.1007/978-981-19-4044-6_14

they are slow, inconsistent, and prone to human mistakes. The form and length of the turbines also makes bare-eye examination difficult [7–9].

Related Works

The goal of this research is to develop a foundation for detecting and classifying flaws on the surfaces of turbines to use a machine learning technique. In recent times, the visual surface examination is performed on a wide range of items, including front seals and PU packaging. Several fault detections, segmentation solutions are proposed and performed successfully.

Depending on the requirements, there seem to be a variety of characteristics to be picked. It is discovered that characteristics based on texture descriptions produced superior outcomes. Techniques for automatically selecting characteristics are not frequently employed. Features extraction strategies are used by a community of scholars. There seem to be numerous classifications accessible to identify faults [10]. Many researchers [11] found that the Support Vector Machine (SVM) used to have a high degree of accuracy. Some researchers have discovered Artificial Neural Network (ANN) efficiently to be satisfactory for this kind of purpose. It has been noted that limited study on categorizing faults on the surfaces of turbines is published. This research proposes and implements a methodology for recognizing and categorizing surface structure in the turbine at the laboratory scale. Surfaces images showing the damaged areas of 10 utilized turbines are taken using a Charge Coupled Device (CCD) sensor and a small microscope probing in a non-destructive method. The first step is to classify the problems into three groups: typical wearing, no-defect, and also the remainder of the faults. The flaws are again divided and also faulty regions determined to use the expanded maximum transformation. For categorization, many textural characteristics depending on histogram and GLCM are retrieved from the segmentation images. Different classifications like Bayes, k-NN, ANN, and SVM are used to test the remaining flaws. All of these processes are performed at various levels, with the optimum scale selected based on segmented and higher accuracy. The computing effectiveness of the whole inspection process is ensured by Gaussian scaled separation approach.

Proposed Works

The inner surface image capture of the utilized cylinders is done with a CCD camera and a small optical coupler. Figure 13.1 shows a schematic representation of the experimental set-up utilized for images acquired. The lighting is provided by an LED that would be dispersed by a reflected ray. The light's course is depicted in the schematic image [12]. The camera, which is coupled to an optical probing and illuminated, maybe rotated horizontally and vertically within the turbine to take

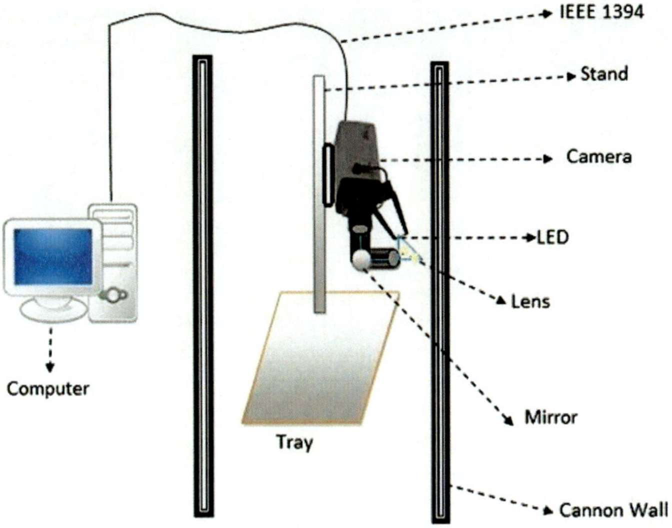

Fig. 13.1 Schematic diagram of the experimental set-up

photos of the whole inside area. The device is calibrated by recording the image of an optic grating with a resolution of 1/100 mm and calculating the pixel that correspond to it.

A total of 1000 images is chosen from a massive pool of images collected from the inner side of 10 utilized turbines at specific segments, with 200 images per class, including such eroding, corrosive pits, generally, corrosion, and also no. Figure 13.2 depicts the 5 types of faults. Cross-validation uses half of the photos from every category to evaluate the accuracy of segments and classification. Depending on the actual truth, the square of the exposed surfaces in metric measurements is estimated and validated for correctness. By physically segregating the images utilizing Adobe Photoshop CS5 program and turning the faulty pixel to red and also the remainder pixel to blues, the high accuracy of the imperfection is produced. MATLAB is used to award 1 and 0 to defective and non-defective squares, accordingly, depending on the manual choice of colors.

Results and Discussions

To use the forward's feature selection algorithm, a few essential parts for categorization are selected automatically from the image characteristics. Several classifiers, including Mahalanobis distance matrix classifiers, Euclidean minimum distance classification algorithm, Naive Bayes-Nearest Neighbor (k-NN), Artificial Neural Network (ANN), and SVM Classifiers, are trained using the relevant attributes acquired via selecting features (SVM). The clustering algorithm for each of them is

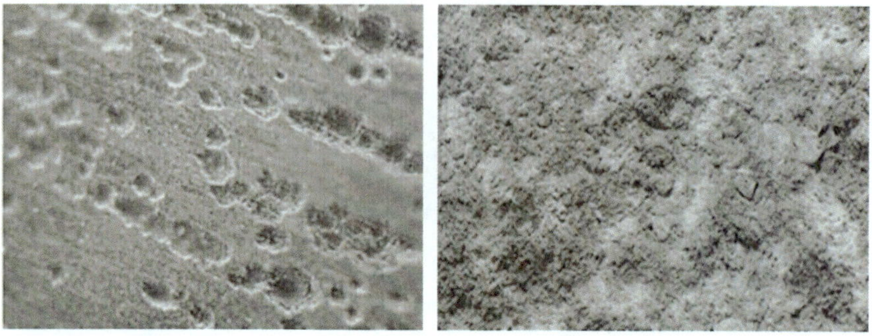

Fig. 13.2 Dataset's categories

used to evaluate their accuracy. A flow diagram depicting the steps taken to finish the process categorization is presented in Fig. 13.3. All these operations are performed out on six different image numerical strengths, with the optimal scaling being chosen depending on segment and accuracy rate. This spatial pyramid idea is developed to ensure that the research' computing effectiveness never is compromised at the expense of efficiency.

The information is classified into three groups for the classification step, including 200 images representing regular wear, 200 images representing no defect, and 600 images representing the other three types of errors. Half of every class is allocated for programs of study. Cross-validation is used to divide the training and certification in 50% of the time. From every scale, two characteristics are derived for the classification step: average gray level and sample variance. After using the Otsu method

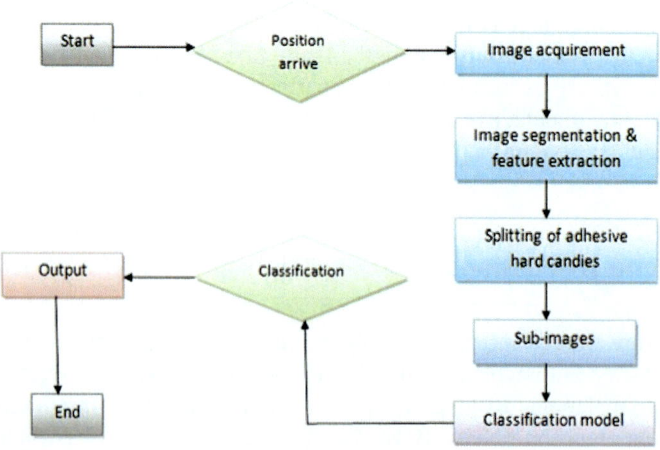

Fig. 13.3 Process of actual fault

Table 13.1 Classification

	Others	No-defect	Normal wear
Others	597	0	5
No-defect	0	200	0
Normal Wear	5	0	197
Accuracy = 99.3%			

to global threshold the photos, the average gray value is determined. Because of its high brightness in any of those photos, the average gray level of typical wearing is considered lower than some other problems. Due to the obvious homogeneity in gray levels, the standard error for images without any defects would be lower. For every level, Mahalanobis and Euclidean minimal similarity classifier is a learning guide depending on these characteristics. Both classifications exhibit nearly comparable levels of accuracy.

The Euclidean minimum distance learner is chosen because it takes less time to compute than the Distance measure classifier, which requires the creation of a covariance matrix. Table 13.1 shows the prediction model for the first classification employed in this study. Following the initial categorization, the photos having typical wearing are enhanced to make the fault appear darker throughout all types of faults. Several adaptive threshold approaches are investigated, including Otsu, area dividing, watershed transformation, and expanded maxima convert.

To reduce sounds, morphology treatments including image erosion and dilation are used following classification. The correctness of the split area is checked against the previously determined underlying data. For every fault class, Table 13.2 lists the procedures and their reliability. Figure 13.4 presents a comparison of several segment findings for a typical wear image. Including all types of flaws, the expanded maximum transformation is found to provide reasonable performance.

The correctness of the testing data is assessed using Bayes, k-Nearest Neighbor, Artificial Neural Network, and Support Vector Machine classifications. For Bayes, the previous likelihood is considered to be the same for all categories. In the k-NN approach, the number of clusters is adjusted, and also the best k is determined to be 5 depending on assessment efficiency. The variance in assessment and learning efficiency as the number of clusters increases is seen in Fig. 13.5a. For ANN, the number of layers varies, and three hidden levels are chosen for term accuracy. Figure 13.5b shows the change in efficiency as a function of the number of hidden nodes.

The categorization is completed at all scale-space regions. Figure 13.6 shows the changes in reliability for classification phase, segment, and end categorization using

Table 13.2 Various segmentation techniques

Method	Accuracy (%)
Otsu	56
Region based	66
Watershed	82
Extended maxima	87

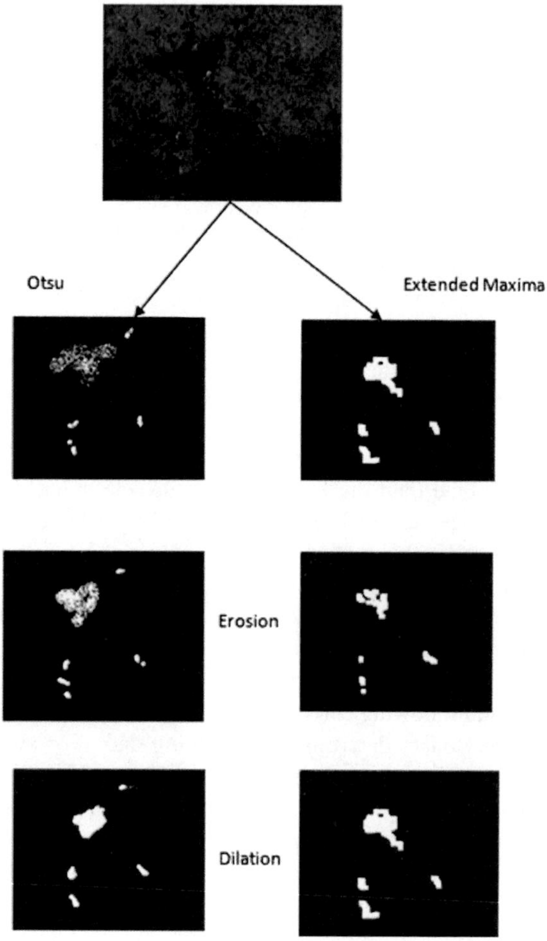

Fig. 13.4 Feature extraction using proposed system

varied scale-spaces of images. The reliability of the initialization step, segmented, and end classification techniques are seen to be preserved until the Gaussian pyramid scaling grade of 3 is reached. The use of scaling separation in pattern categorization showed to be useful in terms of computational power.

The highest image testing period is 0.7 s, a 20% decrease in duration due to lower scalability, and hence it may be used in live time for automated processes.

Conclusion

Surfaces flaws of the used turbine are identified and classified using artificial intelligence and machine learning approaches. Surface area images of a turbine having

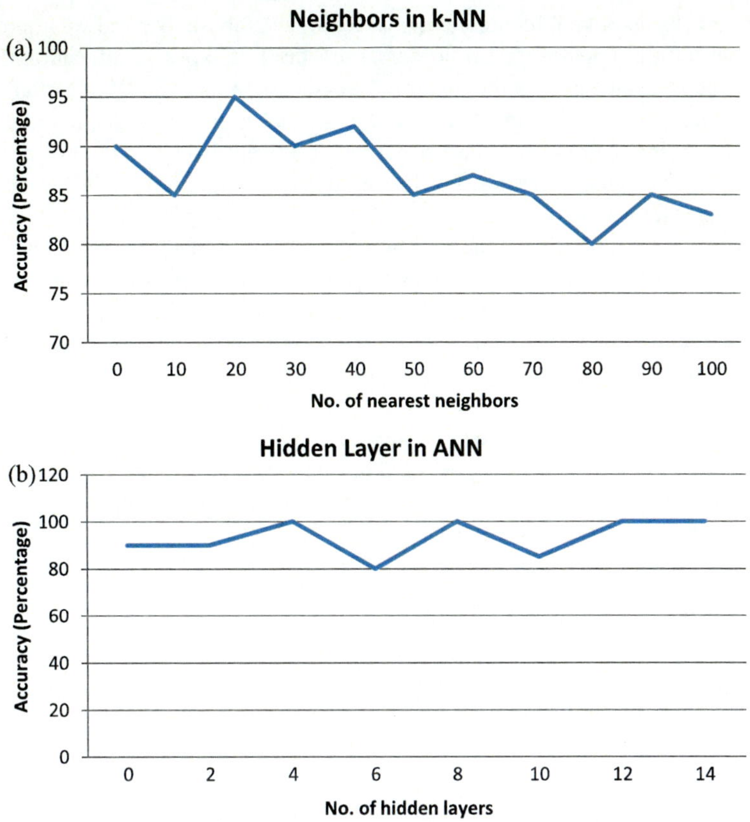

Fig. 13.5 Various classifiers using KNN and ANN

Fig. 13.6 Comparison of proposed and existing system

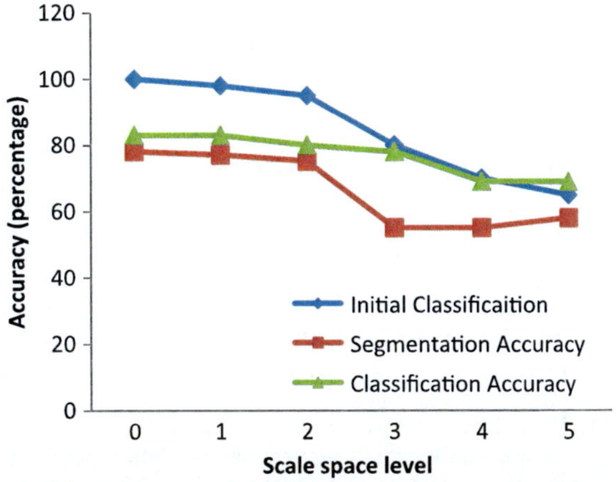

surface cracks are taken, identified, and saved in a database. Natural wear, erosion, corrosive pitting, corrosion, and no-defect are the five types of allocation. Every image type is created as a multiresolution Gaussian pyramid. Unlike other faults, the flaws become larger in normal wear. As a result, an initial classification is carried out to determine the reason for segmentation. The standard error and average gray level are used as characteristics. For this initialization step, the Euclidean distance measure predictor scored well in terms of effectiveness and processing time. Among the various segmentation techniques employed, the extended-maxima transformation worked well. The size of every flaw is quantified metric measurements to use an optic grating to calibrate the sensor. The extracted features yielded 12 texture factors based on the distribution and GLCM. For the classification purpose, five characteristics are chosen to use a progressive forwards feature selection algorithm. Including an efficiency of 96.67%, SVM appears as the top predictor among the various classifications evaluated. The six-level, multi-level Gaussian gradient is developed for computational detection and segmentation of flaws while keeping the same rate of precision. This is found to be well into the image space using grade 3.

References

1. Zhang, W., Tan, A., Zhou, G., Chen, A., Li, M., Chen, X., Hu, Y., et al.: A method for classifying citrus surface defects based on machine vision. J. Food Meas. Charact. **15**(3), 2877–2888 (2021)
2. Sampath, V., Maurtua, I., Aguilar Martín, J.J., Gutierrez, A.: A survey on generative adversarial networks for imbalance problems in computer vision tasks. J. Big Data **8**(1), 1–59 (2021)
3. Gupta, K.K., Mukhopadhyay, T., Roy, A., Roy, L., Dey, S.: Sparse machine learning assisted deep computational insights on the mechanical properties of graphene with intrinsic defects and doping. J. Phys. Chem. Solids **155**, 110111 (2021)
4. Xu, C., Zhu, G.: Intelligent manufacturing lie group machine learning: real-time and efficient inspection system based on fog computing. J. Intell. Manuf. **32**(1), 237–249 (2021)
5. Stephen, O., Maduh, U.J., Sain, M.: A machine learning method for detection of surface defects on ceramic tiles using convolutional neural networks. Electronics **11**(1), 55 (2021)
6. Dong, C.Z., Catbas, F.N.: A review of computer vision-based structural health monitoring at local and global levels. Struct. Health Monit. **20**(2), 692–743 (2021)
7. Benbarrad, T., Salhaoui, M., Kenitar, S.B., Arioua, M.: Intelligent machine vision model for defective product inspection based on machine learning. J. Sens. Actuator Netw. **10**(1), 7 (2021)
8. Chen, L., Yao, X., Xu, P., Moon, S.K., Bi, G.: Rapid surface defect identification for additive manufacturing with in-situ point cloud processing and machine learning. Virtual Phys. Prototyping **16**(1), 50–67 (2021)
9. Yasuda, T., Ookawara, S., Yoshikawa, S., Matsumoto, H.: Machine learning and data-driven characterization framework for porous materials: permeability prediction and channeling defect detection. Chem. Eng. J. **420**, 130069 (2021)
10. Latchoumi, T.P., Parthiban, L.: Quasi oppositional dragonfly algorithm for load balancing in cloud computing environment. Wirel. Pers. Commun. 1–18 (2021)
11. Hu, W., Wang, W., Ai, C., Wang, J., Wang, W., Meng, X., Qiu, S., et al.: Machine vision-based surface cracks analysis for transportation infrastructure. Autom. Constr. **132**, 103973 (2021)
12. Voronin, V., Sizyakin, R., Zhdanova, M., Semenishchev, E., Bezuglov, D., Zelemskii, A.: Automated visual inspection of fabric images using a deep learning approach for defect detection. In: Automated Visual Inspection and Machine Vision IV, vol. 11787, p. 117870P. International Society for Optics and Photonics (2021, June)

Chapter 14
High-Performance Computing Framework for Virtual Memory Using CNN

K. Rameshwaraiah, S. Sree Hari Raju, and K. Ashok Kumar

Introduction

Machine learning has shown impressive results in numerous image processing applications identification, natural language processing, text analytics, and safety in recent machine learning research [1–3]. With larger deep neural networks and also more training information, highly accurate models are created. When the scale of the deep learning models and also the number of training examples are multiplied, the quantity of computing needed increases proportionally [4]. It takes time to train convolutional neural networks on a computer due to the higher quantity of machine learning to compute; therefore, dispersed and great-throughput computing structures & capabilities were necessary for producing period machine techniques using great results [5]. Remote computational intelligence systems have explored requirements specifications. Operators using decentralized deeper learning technologies construct deep convolutional neural networks in a decentralized manner. Furthermore, participants must share vast supervised neural variables themselves, which ultimately generate significant communications latency, accounting significant fraction of the total time spent on dispersed machine learning [6]. As a result, the idle period of processing elements including the CPU and GPU increases, resulting in a decrease in computing resource usage [7].

K. Rameshwaraiah (✉) · S. Sree Hari Raju · K. Ashok Kumar
Nalla Narasimha Reddy Education Society's Group of Institutions, Hyderabad, Telangana, India
e-mail: ramhyd20@gmail.com

S. Sree Hari Raju
e-mail: rvs2raju@gmail.com

K. Ashok Kumar
e-mail: ashok.kondra123@gmail.com

Ch. Satyanarayana et al. (eds.), *Proceedings of the International Conference on Computer Vision, High Performance Computing, Smart Devices and Networks*, Advanced Technologies and Societal Change,
https://doi.org/10.1007/978-981-19-4044-6_15

The significant communications cost is caused by network processing and also the huge and dominant variable transmission [8]. Storage copying procedures are performed during the calculation, and the associated outgoings also are incurred when working using a multi-layer communication protocol. The number of variables in a deeper learning model grows rapidly as its size grows. For quick and huge parallel computing, ring interconnection connectivity is required. Humans offer a strategy for effectively reducing the communication cost of huge, dispersed machine learning [9].

Related Works

Our technique does this by allowing remote employees to quickly exchange supervised neural variation. The sharing process can be accomplished by creating a new mind network environment and granting access to the system storage to all employees [10]. The proposed global supervised neural main memory architecture could be used in high-performance groups coupled using high-speed networking techniques [11]. Moreover, the proposed architecture employs the RDMA technology, which avoids data communication transfer procedures across implementation and kernel-level memory buffers [12]. Finally, employing RDMA for reading/writing information from remote node storage, the proposed scheme provides a methodology for sharing supervised learning variables between dispersed employees.

Hogwild proved that machine learning activity using the same weights and separate information shards may learn DNN using asynchronous SGDs in the main storage structure without locked, which is among the globally computational intelligence optimization techniques depending on Stochastic Gradient Descent (SGD). In particular, the Dogwild architecture, which is a Hogwild modification, conducts DNN learning by asynchronously swapping weight and biases between both the masters and slave operations [13]. Whenever the master process gets varied from the slave operations, it changes the world values and distributes the revised values to every slave program at the same time. Every slave process of understanding the grades to the masters after updating a most recently obtained weight with the gradient it has discovered on its own. As a method of asynchronous SGD, the Dist Belief proposed approach Downpour SGD, which enables a huge number of modeling replicates. The Sandblaster batching optimization technique, which is a system that supports diverse networked systems, is also developed. Inside an asynchronously SGD, every DNN trained agent could acquire parameters at varying speeds without synchronization cost, maximizing the use of CPU & internet connectivity. The capabilities for a heterogeneous HPC method, which consists of CPUs & GPUs with varying requirements (clock speed, number of units, so on), could be effectively utilized to achieve optimum computing capacity.

Proposed Method

The goal of parallel processing, machine learning is to use increased technology nodes to train enormous data and huge DNN algorithms. The needed transmission power grows quickly for the size of DNN & also a dimensionality (i.e. Resolution in image information) for generating accuracy grows. As a result, low-bandwidth systems like 1Gbps Ethernet are unable to cope using the fast-growing internode supervised neural variable traffic. As a result, many businesses and research rely on high-speed network interfaces, which are commonly found in powerful computers and powerful computing platforms.

The HPC system proposed in the work comprises numerous processing elements of multiple GPUs, a different storage node with enough storage space to offer distant memory space, and also an Infiniband switching that combines those components. SMU Librarian seems to be a consumer stationary framework that offers application software for distributed software procedures and is statistically connected with the procedures during program execution. The SMU Device Controller and Infiniband Communication Layer were also kernel-level components that must be activated before the recommended process can be executed. The storage space of the nodes is provided by the SMU Servers by pooling it in preparation, or it could be provided on-demand for customer applications. The provided storage sectors of customer activities should indeed be identified in the hosting channel adaptor drivers to facilitate RDMA operations from the deeper active learning in the compute nodes. The virtual to physical address translation data for the specified web page are enrolled throughout the Signup process. Pooling storage approaches have often been favored because of the longer registration time. The SMU Device Controller seems to be a kernel-level device driver which translates a virtual address of the DNN learning cycle to the actual memory space sectors that act as the caching for the distant sharing program memory, as shown in Fig. 14.1.

Results and Discussions

The structure of the SMU Client, including its recollection pool management technique for providing the main memory buffering, can be seen in Fig. 14.2. The SMU Server is used as a user program on Linux. The multi-threaded design of the SMU Server allows it to spin to process requests from numerous SMU Peripheral Devices at the same time. The messaging reception threading temporarily stores the response message being sent by SMU Peripheral Devices in the messaging rings. The messaging recipient process wakes all the demand processes in the memory pool followed by information collection in the rings. Perhaps one of those, in a circular queuing pattern, accepts one request message from the messaging rings, analyzes it, and would then return to the ready queue whenever the messaging process is complete. The SMU Server uses the buddy memory control method, which would be

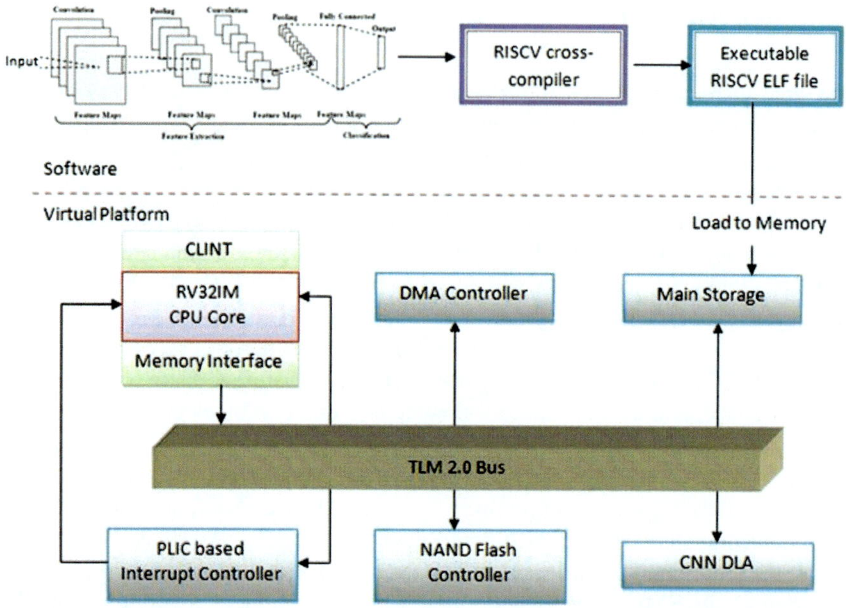

Fig. 14.1 System architecture

typically used for normal pages assignment in a Linux kernel, to handle given information as a pool. With less than 4 KB & even a maximum level of 32 GB, a storage block length could be customized, a maximum length can also be changed. Libibverbs have been used to build engaging lessons. The SMU provides the network with the remote management keys needed to the SMU system Driver for readily accessing main storage buffers during an address space creation stage.

The SMU has been used to allocate and deallocate memory space across distributed machine learning employees, as seen in Fig. 14.3. Firstly, the master's deep learning person uses the SMU API to construct a main memory buffer on the SMU Controller. The expert task distributes SHMKey for the storage space for the workers establishing it. Additional workers which would like to be allotted a main storage buffer produced by an expert employee need to have their channel of communication. The person who gets the SHMKey from the supervisory employee transmits the SHMKey including a distributed memory allocation application. If the employee asks for the same sized sharing memory data allocation and delivers the very same SHMKey, the SMU Server uses the Encryption Key for the sharing buffer cache. After the assignment operation has finished, the SMU User's shared memory buffer could be distributed across the dispersed deep learning workers.

Table 14.1 displays the variable sizes and computation times of six CNN models that are assessed. Because batch length represents several images to be learned, changing the batch size affects the computational effort. Caffe needs substantial memory resources as the batch size grows, hence there seems to be a maximum

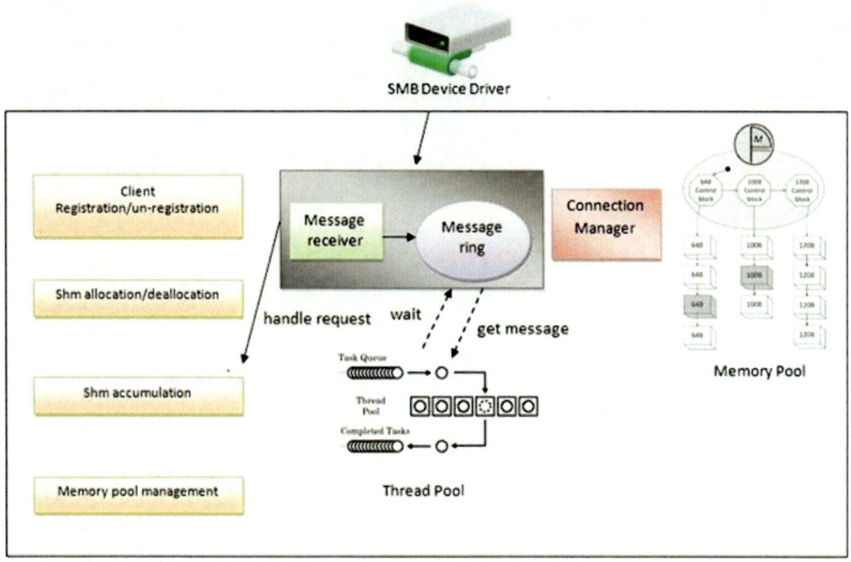

Fig. 14.2 Architecture of the SMU server

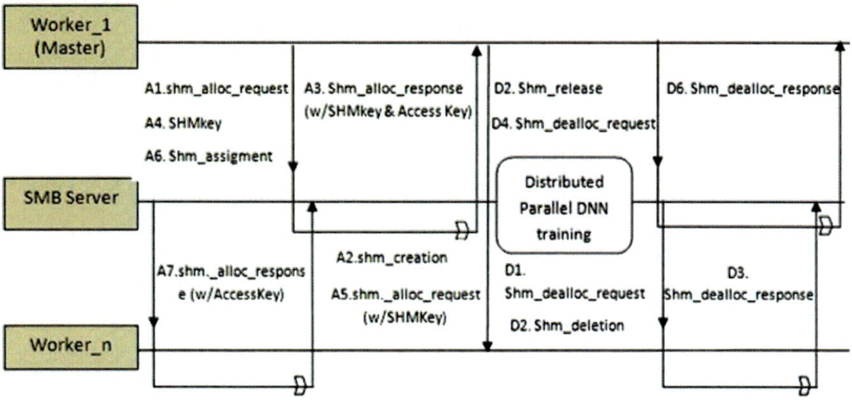

Fig. 14.3 Observational research on sharing data storage

batch length at which Caffe could train every model in an execution. In terms of variable length and computational cost, the proposed CNN models have a wide range of values. The time it would take to compute a variable is never dependent on the magnitude of the variable. The VGG16 contains the most parameters (140 million), yet it takes less time to compute than the other four methods.

In the parallel situation, the connection times of MPI and SMU are compared in Figs. 14.4 and 14.5. During every iterative process of DNN coaching, the sequences situation does not consider data processing overfitting. As demonstrated in Fig. 14.4,

Table 14.1 ML performance measures

Network specification	CNN					
	Inception_v1	Inception_v3	Resnet_50	Inception_resnet_v2	Resnet_152	VGG16
Variable (million)	13.5	24.6	31.9	57.2	66.6	139.1
Variable size (Mb)	51.21	94.32	121.14	214.61	253.64	527.26
Compute time (ms)	190	451	234	295	219	191
Batch size	51	33	19	7	8	65

as the model size & number for staff increased, an MPI product's transmission rate increased rapidly, whereas the SMU product's transmission rate gradually increases. Figure 14.4 depicts general trends, while Fig. 14.5 compares the MPI and SMU techniques by adjusting the amount of machine learning workers for each of the six deep learning techniques. In all scenarios, it has been shown that the SMU technique is 1.1 to 4.2 times faster than the MPI approach.

For any dataset and any number of staff, the SMU work demonstrated the MPI approach. The SMU technique has a transmission time of 0 ms up to 24 employees in an Inception v1 using a very tiny design capacity, & 45 ms, even the number of staff is increased to 32, which would be 7 times faster than the MPI approach. In the instance of Inception v3, which has a shorter variable size and a longer computational cost, the calculation time hides the entire calculation time of the SMU technique. However, whenever the number of processes is increased to 32, the MPI product's transmission rate exceeds the computational effort by 2.3 times. The variation in

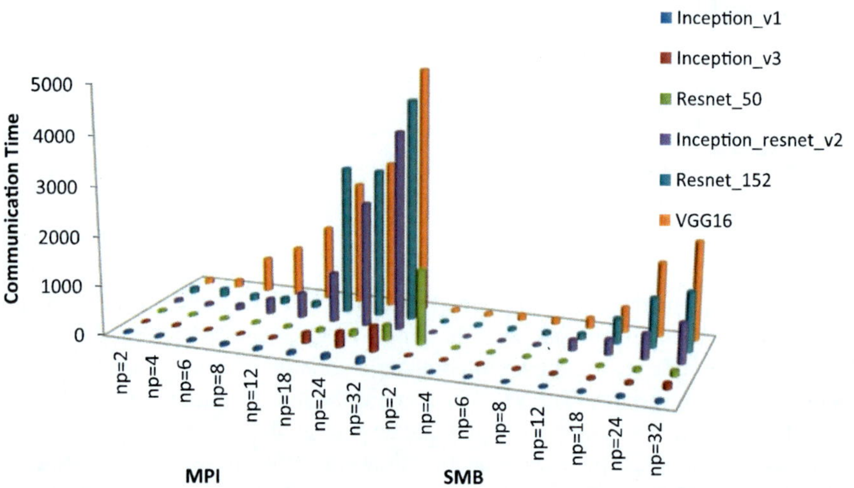

Fig. 14.4 Comparison of MPI-based and SMU

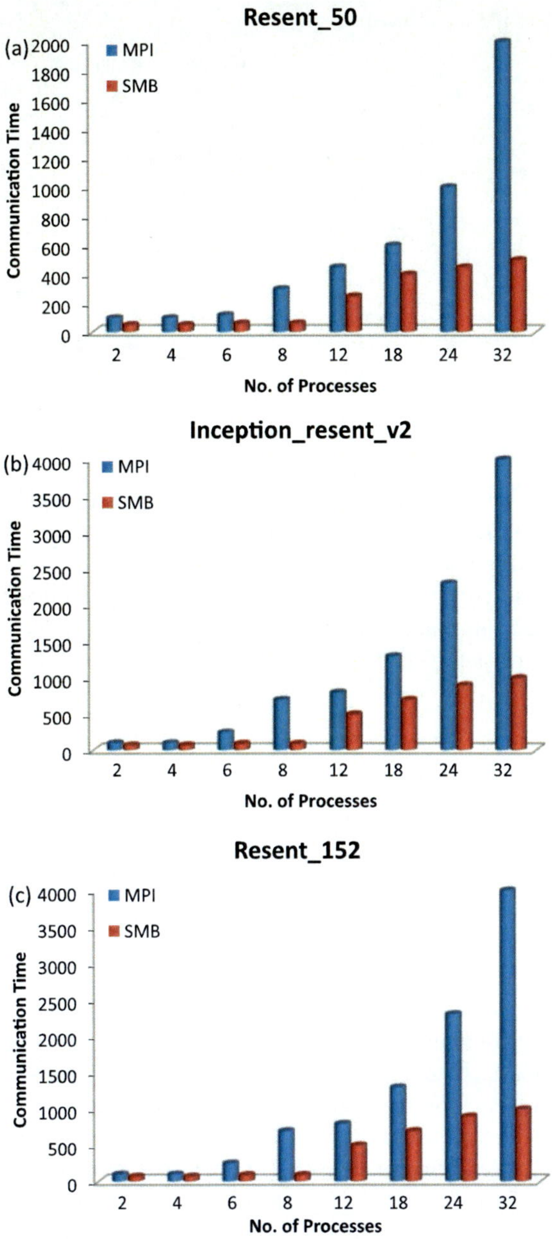

Fig. 14.5 Comparison of transmission times for each type

Fig. 14.5 (continued)

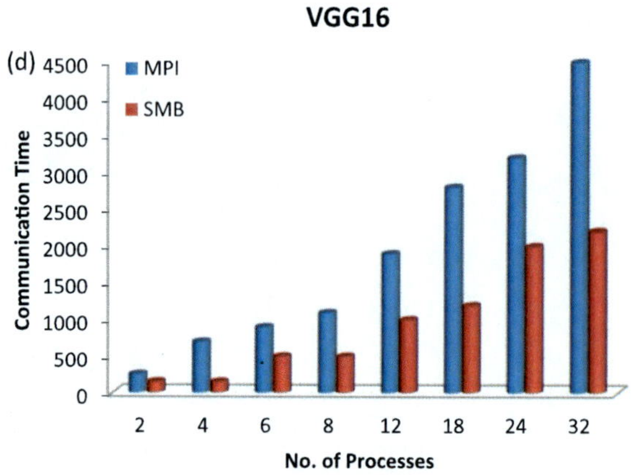

the true communication gap between the MPI & SMU system reduces as a model size rises, but in the particular instance of Resnet 50, 4.8% stronger in an Original conception Resnet v2 framework, & 2.0 times higher in a Resnet 152 & VGG16 prototype, the SMU technique is 5.2 times faster than the MPI technique. The SMU-based machine learning system's average of 1 repetition number of training has increased by 3.5 times.

Conclusion

The proposed SMU makes use of media transmission memory to transport variables from the memory of the local computer to the memory of the remotely distributed node, and interchange DNN variables by learning to write the remote distributed memory buffer. Storage duplication and internetwork execution have a significant influence on communication latency. The proposed SMU-based system was 2.1 times quicker than the MPI-based system in a sequential situation for asynchronous SGD. The communications with the SMU-based dispersed DNN training platform are two to seven times more efficient than using an MPI-based system in a concurrent scenario of asynchronous SGD, where computational resources overlapped. Only a minimal predictive performance is validated because the results of these experiments are achieved using a single SMU Server. The amount of SMU Server, on the other hand, could be raised to boost efficiency, and also capacity could be calculated according to the capacity of models & also the number of supervised neural employees. The suggested SMU-based method lays the groundwork for expediting the networked analysis of information DNN learning in a powerful computational environment along with high speed and accuracy.

References

1. Zhang, X., Zhang, T., Lu, J., Fu, X., Reveriano, F.: The effect of high-performance computer on deep neural network. Eng. Sci. **15**, 67–79 (2021)
2. Singh, A., Prakash, S., Kumar, A., Kumar, D.: A proficient approach for face detection and recognition using machine learning and high-performance computing. Concurrency and Comput.: Practice Exper. **34**(3), e6582 (2022)
3. Haseeb, M., Saeed, F.: High-performance computing framework for tera-scale database search of mass spectrometry data. Nature Computat. Sci. **1**(8), 550–561 (2021)
4. Jin, S., Li, G., Song, S. L., Tao, D.: A novel memory-efficient deep learning training framework via error-bounded lossy compression. In: Proceedings of the 26th ACM SIGPLAN Symposium on Principles and Practice of Parallel Programming, February, pp. 485–487. (2021)
5. Lima, A.L.D.C.D., Aranha, V.M., Carvalho, C.J.D.L., Nascimento, E.G.S.: Smart predictive maintenance for high-performance computing systems: a literature review. J. Supercomput. **77**(11), 13494–13513 (2021)
6. Deng, C., Sui, Y., Liao, S., Qian, X., Yuan, B.: GoSPA: an energy-efficient high-performance globally optimized sparse convolutional neural network accelerator. In: 2021 ACM/IEEE 48th Annual International Symposium on Computer Architecture (CA), June, pp. 1110–1123. IEEE (2021)
7. Pandey, S., Nagwani, N.K., Verma, S.: Aspects of programming for the implementation of convolutional neural networks on multisystem HPC architectures. J. Phys. : Conf. Series **2062**(1), 012016 (2021). IOP Publishing
8. Patel, S., Liu, T., Guan, H.: free lunch: compression-based GPU memory management for convolutional neural networks. In: 2021 IEEE/ACM Workshop on Memory Centric High-Performance Computing (MCHPC), November, pp. 1–8. IEEE (2021)
9. Latchoumi, T.P., Parthiban, L.: Quasi oppositional dragonfly algorithm for load balancing in cloud computing environment. Wireless Personal Commun. 1–18 (2021)
10. Yao, C., Liu, W., Tang, W., Hu, S.: EA: energy-aware adaptive scheduling for CNN inference on high-performance GPUs. Future Generation Comput. Syst. (2022)
11. Balamurugan, K.: Metrological changes in surface profile, chip, and temperature on end milling of M2HSS die steel. Int. J. Mach. Mach. Mater. **22**(6), 443–453 (2020)
12. Chang, S.E., Li, Y., Sun, M., Shi, R., So, H.K.H., Qian, X., ..., Lin, X.: Mix and match: A novel FPGA-centric deep neural network quantization framework. In: 2021 IEEE International Symposium on High-Performance Computer Architecture (HPCA), February, pp. 208–220. IEEE (2021)
13. More, N., Galphade, M., Nikam, V. B., Banerjee, B.: High-performance computing: a deep learning perspective. In: Deep Learning and Edge Computing Solutions for High-Performance Computing, pp. 247–268. Springer, Cham (2021)

Chapter 15
High-Performance Computing Center Framework for Smart Farming

Chandra Sekhar Akula, Venkateswarlu Sunkari, and Ch. Prathima

Introduction

In past times, increasing agricultural productivity has received a lot of attention. The production forecasts are unsustainable since the world population is increasing uncontrollably, while agricultural lands are decreasing dramatically [1]. By 2050, worldwide grain yields would have to treble to fulfill the needs of ten billion people [2–4]. Smart agriculture is necessary to boost agricultural output due to the challenges of expanding the total cultivated area [5]. Smart agriculture, also known as precision agriculture, is a fascinating idea that uses cutting-edge technologies such as remotely sensed, geographic information systems, and telecommunications, including powerful computational mechanisms [6] to monitor, analyze, as well as respond to agricultural variations.

Raising agricultural output has garnered considerable attention. The output predictions are unachievable since the global population is uncontrollably expanding whereas agricultural lands are rapidly diminishing. To meet the demands of 10 billion people by 2050, global grain production is to be tripled. Intelligent farming is required to increase farm productivity to overcome the difficulties of growing total cropped area. Sustainable farming, also known as precision agriculture, is an intriguing

C. S. Akula
Avanthi Institute of Engineering and Technology, Vizianagaram District, India
e-mail: chanduforu@gmail.com

V. Sunkari
Addis Ababa University, Addis Ababa, Ethiopia

Ch. Prathima (✉)
School of Computing, Mohan Babu University, Tirupathi, India
e-mail: chilukuriprathi@gmail.com

Ch. Satyanarayana et al. (eds.), *Proceedings of the International Conference on Computer Vision, High Performance Computing, Smart Devices and Networks*, Advanced Technologies and Societal Change,
https://doi.org/10.1007/978-981-19-4044-6_16

concept that employs cutting-edge innovations like aerial imagery, geographic information systems, mobile communications, and computationally intensive methods to supervise, analyze, and respond to farmland differences.

Related Works

Deep learning, which takes advantage of the superpowers of elevated computer networks, has recently emerged as a promising method for solving challenges concerning image categorization reliability. Deep learning in neural networks, often known as deep neural networks, has received awards in classification techniques and machine learning competitions [7–9]. DNNs feature several computing stages with sophisticated systems to mimic sophisticated as well as imperfect actual information, resulting in a high computational burden [10]. As a result, HPC platforms must integrate DNNs for them to be a strong and valuable technology. DNNs have already been utilized in PA only seldom due to the stringent hardware requirements.

In light of this pressing need, the focus is to research using modern methods like DNNs and UAVs to analyze the condition of Indian paddy fields [11]. The overall thickness of paddy fields, comprising sparse and standard concentration, is used to determine the grade. It is worth noting that paddy concentration is a key component in determining whether or not a harvest will be effective [12, 13]. Researchers substantial percentage images of the paddy fields from low elevations using unmanned aerial vehicles (UAVs). Next, to examine the performance of the paddy fields, DNNs are utilized to categorize the image. Testing is carried out using high-performance computing (HPC) systems comprised of numerous powerful computers. They chose 10-day-old rice paddies in Deltaregions as a research study for an image collection.

Proposed Method

Figure 15.1 depicts the theoretical method for analyzing the sustainability of rice crops using sophisticated approaches. Deep neural networks are used to identify the remaining images after learning the collected data. They give producers an important approach that is adopted in the categorization findings of deep learning-based processors. Owing to the increased computational burden of deep learning techniques, an excellent performance computer machine is used to train the deep learning-based classification.

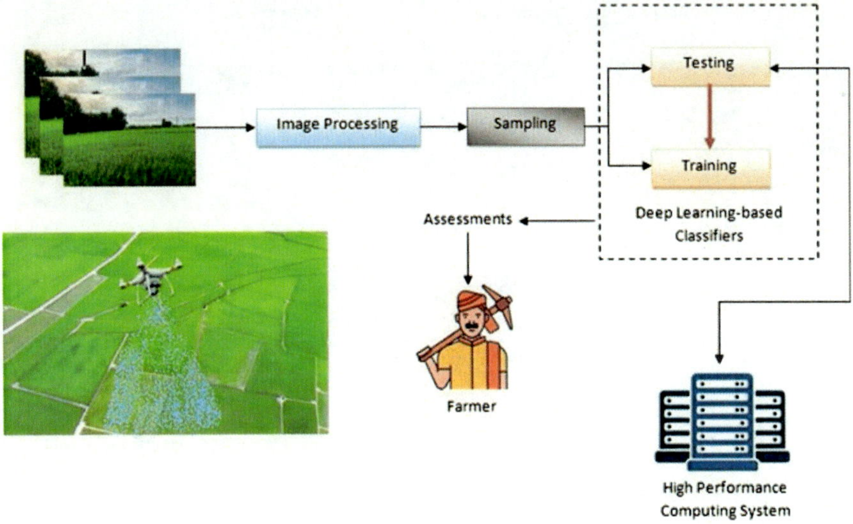

Fig. 15.1 Work flow of the proposed system

Results and Discussion

The photos must also include the entire agricultural area, which is often tiny and irregular in shape in India. Aerial images are undesirable, as per the rigorous criteria because the grade is highly dependent on weather, clouds condition, poor sharpness, and low processing. Furthermore, aerial images are extremely costly and they can be obtained at any moment. Drones/UAVs are employed in this investigation to collect images from moderate and extremely low altitudes. Drone/UAV photos obtained from a low altitude have various advantages, including minimal price, image quality, high quality/bit-rate, noise avoidance, and so on. Furthermore, users could quickly swap out camera angles varieties to collect numerous groups in response to a wide range of goals.

Image enhancement is a crucial component that determines the performance of any algorithm. Some elements in the external environment, including wind, sunshine, weather, etc. have effects on the effectiveness of photos gathered by UAVs in paddy fields. As a result, the images are thoroughly edited to increase their clarity. Figure 15.2 depicts the four components of image preparation.

A CNN often has multiple stages, which are divided into two categories: Convolutional layers and pooling layers. The huge structure is formed by the alternation of two types of layers. The final tiers, like in typical ANNs, are completely linked tiers with complete links to the prior levels. Figure 15.3 shows a CNN instance, notably LeNet-5. Convolution and image rectification, which corresponds to convolution operation and average pooling, accordingly, are two main phrases that describe the intriguing concept of CNNs. Deep learning in general, including CNN's in specific, have proven

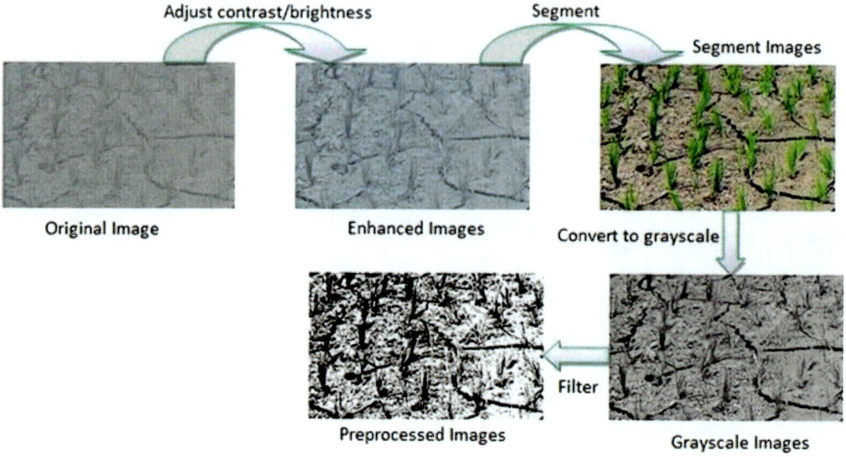

Fig. 15.2 Image preprocessing

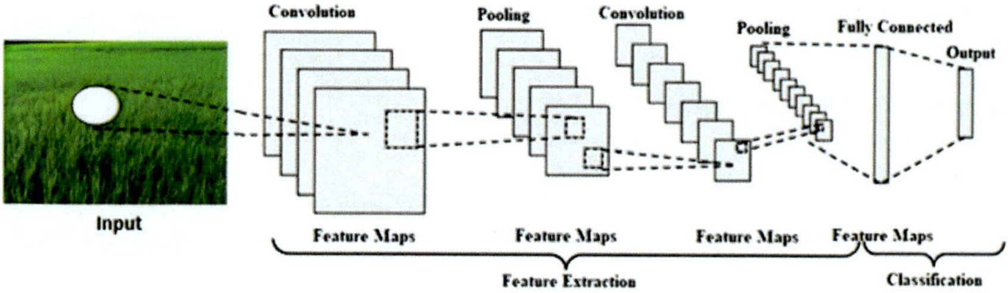

Fig. 15.3 A typical architecture of LeNet

its worth by winning a slew of image categorization tournaments. This is the primary rationale for deciding to use this strategy to solve the image categorization problem.

The footage for this investigation is taken in paddy fields in the Delta regions of southern India. The objective is to rate the performance of 10-day-old paddy fields. The images are taken in two rice paddy fields with a total size of around 2 hectares. The total number of photos obtained is around 800. The authors choose 200 heavily processed photos and categorize them into two classifications: sparse concentration and normal concentration, depending on the skills and experiences of farming professionals. They used 160 photos for learning and 40 images for evaluating CNNs in this dataset.

The results obtained from the deep learning-based classifiers are shown in Table 15.1. Because their quantities are around 0.75, the statistical indicators are very appropriate. As previously stated, the effectiveness of CNNs is determined by their configuration settings. Because DNNs are time-consuming, the trial-and-error technique is not perfect. DNNs require roughly 48 h to converge in one configuration. As

Table 15.1 Performance measures

Training phases		
Names	Values	Formulas
MSE	0.302	$\text{MSE} = \frac{1}{2}\sum_{x=1}^{n}(P_x - O_x)^2$
Testing phases		
Names	*Values*	*Formulas*
Accuracy	0.721	*Accuracy* = (True Positive + True Negative)/(True Positive + True Negative + False Positive + False Negative)
Precision	0.7618	*Precision* = True Positive/(True Positive + False Positive)
Recall	0.7124	*Recall* = True Positive/(True Positive + False Negative)
F1 score	0.741	*F1Score* = 2* True Positive/(2True Positive + False Positive + False Negative)

Fig. 15.4 Sample performance measures over rice-field productivity

a result, several optimization strategies may be used in future research to determine the ideal DNN configuration.

Furthermore, the production of paddy fields is examined using the same paradigm in research consideration. Images of 5-day-to-harvest paddy fields are analyzed to determine production, as shown in Fig. 15.4. The image depicts three distinct sorts of paddy fields: normalcy, wind-caused slumping, and illness infections Fig. 15.4a–c.

Conclusions

To analyze the condition of Indian agricultural areas, this research developed a unique paradigm related to modern methods such as deep learning and surveillance drones. Researchers utilize a database gathered from 10-day-old paddy fields in Deltaregions to test the effectiveness of platforms. The interventions addressed are used to rate the performance of paddy fields using the concentration criteria. The experimental findings show that it is fairly suitable for Indian food practice as the accuracy of the system is around 0.72. The effectiveness of the approach, nevertheless, may be enhanced. Heavily processed photos are relatively huge, and they may include an amount of fuss. DNNs could be acquired more quickly by compressing the relevant details in these photos. Additionally, DNNs' training period is greatly shortened with relatively small photos. In addition, to enhance item edges, standard image processing methods can be implemented.

References

1. Yuan, J., Liu, W., Wang, J., Shi, J., Miao, L.: An efficient framework for data aggregation in smart agriculture. Concurr. Comput. Pract. Exp. **33**(10), e6160 (2021)
2. Senthil Kumar, A., Suresh, G., Lekashri, S., Babu Loganathan, G., Manikandan, R.: Smart agriculture system with E–carbage using IoT. Int. J. Modern Agric. **10**(1), 928–931 (2021)
3. Friha, O., Ferrag, M.A., Shu, L., Maglaras, L.A., Wang, X.: Internet of things for the future of smart agriculture: a comprehensive survey of emerging technologies. IEEE CAA J. Autom. Sinica **8**(4), 718–752 (2021)
4. Sanjeevi, P., Siva Kumar, B., Prasanna, S., Maruthupandi, J., Manikandan, R., Baseera, A.: An ontology-enabled internet of things framework in intelligent agriculture for preventing post-harvest losses. Complex Intell. Syst. **7**(4), 1767–1783 (2021)
5. Kumar, P., Gupta, G.P., Tripathi, R.: PEFL: deep privacy-encoding based federated learning framework for smart agriculture. IEEE Micro. (2021)
6. Prathima, C., Muppalaneni, N.B., Kharade, K.G.: Deduplication of IoT data in cloud storage. In: Satyanarayana, C., Gao, X.Z., Ting, C.Y., Muppalaneni, N.B. (eds.) Machine learning and internet of things for societal issues. Advanced technologies and societal change. Springer (2022)
7. Idoje, G., Dagiuklas, T., Iqbal, M.: Survey for smart farming technologies: challenges and issues. Comput. Electr. Eng. **92**, 107104 (2021)
8. Su, Y., Wang, X.: Innovation of agricultural economic management in the process of constructing smart agriculture by big data. Sustainab. Comput. Inf. Syst. **31**, 100579 (2021)
9. Rouzbahani, H.M., Karimipour, H., Fraser, E., Dehghantanha, A., Duncan, E., Green, A., and Russell, C.: Communication layer security in smart farming: a survey on wireless technologies (2022). arXiv preprint arXiv:2203.06013
10. Latchoumi, T.P., Parthiban, L.: Quasi oppositional dragonfly algorithm for load balancing in cloud computing environment. Wireless Personal Commun. 1–18 (2021)
11. Lin, X., Sun, X., Manogaran, G., Rawal, B.S.: Advanced energy consumption system for smart farms based on reactive energy utilization technologies. Environ. Impact Assess. Rev. **86**, 106496 (2021)
12. Kalyani, Y., Collier, R.: A systematic survey on the role of cloud, fog, and edge computing combination in smart agriculture. Sensors **21**(17), 5922 (2021)
13. Qayyum, T., Trabelsi, Z., Malik, A., K., Hayawi: Trajectory design for UAV-based data collection using clustering model in smart farming. Sensors **22**(1), 37 (2021)

Chapter 16
Generalized and Simulated Architecture for Seamless Experiment Conditions in Cloud Computing

G. Hemanth Kumar, D. R. Kumar Raja, and M. Ramu

Introduction

Cloud computing enables architecture, framework, plus software (application) as solutions, that are provided to users as subscription-based, pay-as-you-go facilities [1]. Clouds aspire to enable next-generation data centers by designing and implementing them as a network of cloud applications, allowed to access and deploy apps on request from everywhere in the globe at affordable prices based on their QoS needs [2–4].

Developers with novel suggestions for future Digital services aren't any longer essential to make a significant capital investment in hardware/software systems, and hr, to implement innovative offerings [5]. It benefits IT organizations by relieving companies of the reduced labor of establishing fundamental hardware/software facilities, focusing attention more on innovations and the production of commercial benefits.

Networking sites, online hosted, content providers, and real-time designed to operate data management are some of the classics and developing Cloud-based technologies [6]. Every one of those applications kinds has its list of expectations for composing, setup, and installation [7]. Taking into account the efficacy of Clouds architecture schedule and distribution policies for various application and services

G. Hemanth Kumar
Assistant Professor, School of ECE, REVA University, Bengaluru, India

D. R. Kumar Raja (✉)
Associate Professor, School of CSE, REVA University, Bengaluru, India
e-mail: kumarraja.raj@gmail.com

M. Ramu
Assistant Professor, Department of CSSE, Sree Vidyanikethan Engineering College, Tirupati, India

Ch. Satyanarayana et al. (eds.), *Proceedings of the International Conference on Computer Vision, High Performance Computing, Smart Devices and Networks*, Advanced Technologies and Societal Change,
https://doi.org/10.1007/978-981-19-4044-6_17

paradigms amid varied demand, energy intensity, and disk usage is a difficult challenge to solve [6]. The utilization of proper testbeds, including Amazon EC2, confines tests to the testbed's scalability and making reproducing outcomes exceptionally difficult, given the circumstances that exist in Network settings are outside the tester's supervision.

Related Works

Another option is to use modeling software, which allows for the evaluation of hypotheses previous to developing software in a system whereby experiments may be replicated [8]. Simulation-based strategies, specifically in the context of Cloud coding, where connectivity to architecture requires real-money financing, provide huge benefits because they enable Cloud customers to check their facilities in a repetitive and an easily controlled atmosphere for ready, and tune performance problems before actually implementing on actual Clouds [9]. Simulated systems, on the supplier side, enable the examination of various resources renting alternatives across varied demand and cost ranges [10]. Such research might assist carriers in lowering the price of resource availability while increasing profitability [11]. Without such frameworks, Clouds users, and services providers, must depend on conceptual and inaccurate judgments, or on trial-and-error procedures, which result in inadequate service efficiency and income-generating [12].

Because none of the existing decentralized network simulation tools provide an atmosphere that can be straightforwardly used against the Cloud technology society, researchers propose CloudSim in this article: a fresh, generalized, and highly configurable computational infrastructure that allows streamlined modeling, virtual world, and experiments of starting to emerge Clouds computing systems and business applications [13, 14]. Scientists and manufacturing engineers could use CloudSim to target specific network engineering problems they wish to examine instead of worrying regarding reduced details concerning Cloud-based infrastructures and services.

Grids have become the foundation for providing excellent services for computation and statistics application areas during the last decades. Various Grid emulators, including GridSim and GangSim, have been developed to aid in the discovery and application of novel Grid components, regulations, including middleware. SimGrid is a standard structure for grid system simulations of application components. GangSim is a Grid integrated suite for simulating Grid-based virtualized groups including services [15]. GridSim, on the other extreme, is a modeling toolset for heterogeneity Grid resources that are occurring. Grid objects, people, computers, and networks, and traffic patterns, are all supported. Even though the abovementioned toolchains can simulate and simulate Grid program behaviors in a distributed world with various Grid groups, none of them can meet the architecture and application-level needs that come with the Cloud computing technology. Existing Electrical simulations toolchains, for instance, provide little or no assistance for simulating on-demand virtualized assisted resources and applications administration. Furthermore, Clouds claim to supply services to Cloud users on a subscription-based model under a pay-as-you-go approach.

Proposed System

This integrated development of the CloudSim programming interface including architectural features is shown in Fig. 16.1. The SimJava simulation models engines are the lightest element, and it provides the essential features necessary for higher-level simulations platforms, such as incident buffering and handling, complete system generation, constituent communications, including simulated field position. The GridSim toolkit's modules enable high-level software requirements for simulating numerous Grid topologies, comprising connections including accompanying traffic characteristics, and essential Grid features like commodities, data repositories, work-load records, among information systems. By functionally expanding the key features offered by the GridSim layers, the CloudSim is developed at the next stage. CloudSim adds new features to design and control standardized Cloud-based network infrastructure settings, including expert management interfaces for virtual machines, ram, space, and broadband. During the modeling stage, the CloudSim layers oversee the creation and implementation of core things. That component is effective in deploying and seamlessly administering a large-scale Cloud environment with thousands of network elements at the same time.

This Client Coding layer is the top of the simulated architecture, exposing parameter settings capabilities for hosting VMs, number of customers and applications kinds, including brokers scheduler guidelines. At this level, a Cloud web application may build a mixture of service request percentages, configuration files, and Clouds available circumstances, and run comprehensive tests using the CloudSim's specific Cloud settings. Because cloud technology is such a new field of study, there are few well-defined guidelines, techniques, and methodologies for dealing with the hardware and software system levels difficulties. As a result, there will be a flurry of research activities in business and academia in the coming years to define basic techniques, rules, and software benchmarks predicated on performance circumstances.

Modeling

A Datacenter module for managing customer inquiries models the main hardware facilities connected to Clouds in the simulation. Such demands are for applications pieces that are sandboxed within VMs and require computing resources on the Datacenter's hosting parts. By VM operations, we imply a collection of processes linked to the VM life cycle, such as VM provisioning, VM generation, VM annihilation,

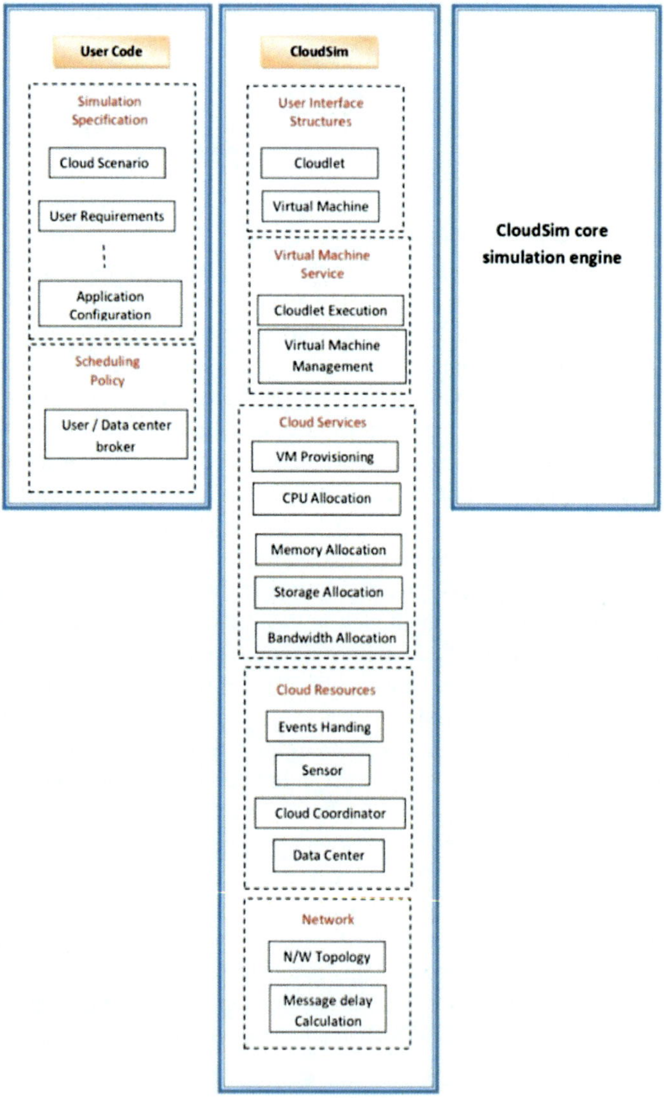

Fig. 16.1 CloudSim architecture

and VM migration. A data center is made up of a collection of hosts that are responsible for managing virtual machines throughout their development cycles. Hosting is a Cloud element that represents a significant computational node, with the preconfigured processor, bandwidth, and transport, and a mechanism for distributing central processing units to virtual servers. This Hosting requirement applies systems that allow single-core and multi-core components to be modeled and simulated.

The distribution of compute nodes to VMs for each Hosted element is based on a host assignment. The strategy considers how many computing cores would be assigned to each VM, and how much of each computing system's performance would be allocated to each VM. As a result, individual CPU cores may be assigned to specific VMs, the capability of a core can be distributed across VMs, cores can be assigned to VMs on demands, and other restrictions can be specified. Each Host reflects the characteristics of a VM scheduling element, which contains guidelines for assigning cores to VMs using space-shared or time-shared rules. The VM scheduling feature may be extended by distributed cloud developers are creating to experiment with more developing countries in particular.

As a result, complex application mappings approaches that allocate particular software parts to compute resources fail to capture the computing generality that is commonly linked with Clouds. Imagine a physical network infrastructure host with a single processor core and a demand to instantiate two VMs on that core at the same time. When both VMs' activities are isolated in practice, the number of resources available toward each VM is limited by the host's overall processing capabilities. This strategic objective should be deemed during both the selection phase, to avoid creating a virtual network that requires more computational power than the host can provide, and during the execution environment, because task measurement is done in each virtual server to share start sharing time slices of the same central entity.

Figure 16.2a depicts a space-sharing strategy for both virtual machines and task modules: because each VM usually requires cores, only one VM may run at any one moment. As a result, VM2 can only be given the core after VM1 has completed the task unit processing. The same is true for activities that are performed within the virtual machine: Because each work item uses just one core, two of these execute at the same time, while the other two were stored until the previous work-groups were completed. Figure 16.2b shows that VMs were allocated using a space-shared strategy, but specific task items inside VMs are allocated using a time-shared approach. As a result, throughout the lifespan of a VM, all work allocated to it periodically environment transition until they are completed. This program allows project units to be planned sooner, but it has a considerable impact on the project duration of work items at the front of the list. In Fig. 16.2c, VMs are scheduled using time-shared planning, whereas work modules were scheduled using space-shared planning. Each VM is given a time-sliced of each processor core in this example, and the slices were subsequently divided into activity groups on a space-shared premise. Because the core is divided, the quantity of computing power accessible to the VM is lower than in the other instances. Because work units distribution is space-shared, each core could only have one job assigned to it, while others are stored for further evaluation. Eventually, in Fig. 16.2d, both VMs and work modules were given a time-shared assignment. As a result, the computing power is shared among the VMs at the same time, and the portions of each VM were split among the workgroups given to each VM at the same time. There are no lines for virtual servers or workgroups in this scenario.

A data center is made up of several hosts that are in charge of maintaining virtual machines throughout their life cycles. The host is a Cloud element that poses a real

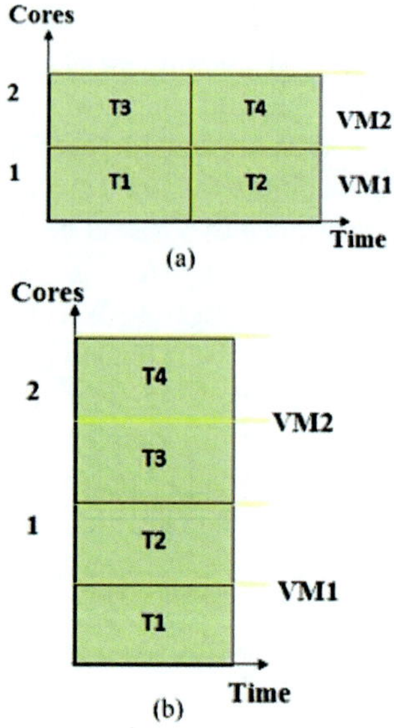

Fig. 16.2 **a** Various scheduling VMs and activities. **b** Space-shared for VM

computational node, with the pre-configured processor, memory, including storing, and a policy for distributing processing cores to virtual servers. The Host component offers platforms that allow single-core and multi-core nodes to be modeled and simulated. The Virtual Machine Provisioner module is responsible for allocating application-specific VMs to Hosts in a Cloud-based datacenter. That module provides researchers with access to a variety of customized methods that help in the design of new VM provisioned strategies depending on efficiency objectives. The VM Provisioner's standard strategy is a basic one that assigns a VM to the Host on a first-come, first-served premise. Such translations are based on system attributes such as the needed amount of processor cores, space, and storage as desired by the Public cloud. Our scientists can write more intricate regulations based on the infrastructure and software requirements.

The communications pattern among fundamental CloudSim elements is depicted in Fig. 16.3. Each Datacenter object records with the CIS Registry at the start of the simulation. CIS offers database-level match-making solutions to connect client requests with appropriate Cloud vendors. Dealers working on behalf of users examine the CIS services for a list of Clouds that provide operational services that meet the needs of the user's applications. If a connection is made, the brokerage installs the program using the Cloud recommended by the CIS. So far, the communications channel has been compared to a simulation research's fundamental process. Based on rules, certain modifications in this process may be allowed. Messaging from Brokerage to Datacenters, for instance, might need some Datacenter approval of the action's implementation, or the largest amount of VMs a customer might generate may be discussed before VM formation.

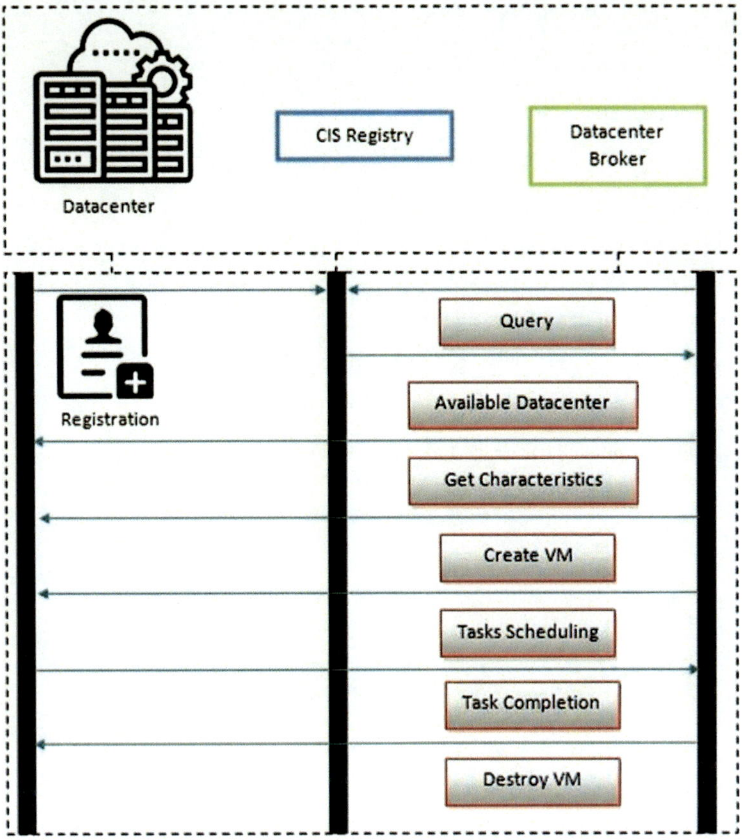

Fig. 16.3 Simulation data flow

Evaluation

Researchers ran several tests to see how much time it took to establish replicated Cloud computing services with a single data center, brokers, and a user. Each study had a different network in the data center, ranging from 100 to 100,000. The purpose of these experiments was to determine the amount of computational power required to set up the Cloud modeling environment, hence the user workloads were ignored. Figures 16.4 and 16.5 show the quantity of days and storage required to run the test as the number of hosts in a data center grows. The need for storage grows linearly, with an operation with 100,000 devices requiring 75 MB of RAM. Although CloudSim storage needs, even for bigger modeled settings, can readily be met by such systems, their model could operate on modest desktops and laptops with low computing capability.

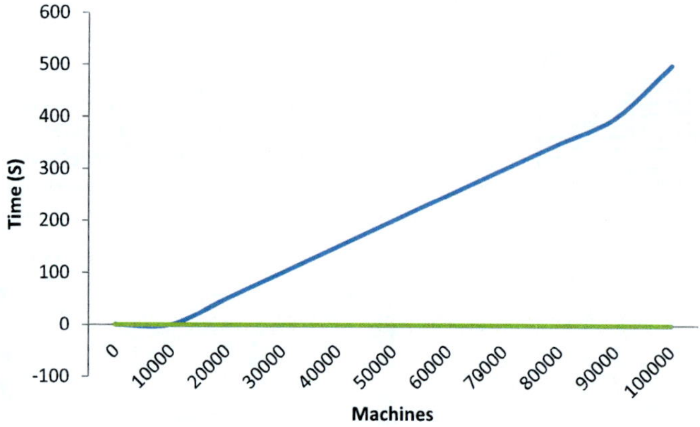

Fig. 16.4 Time to simulate instantiation

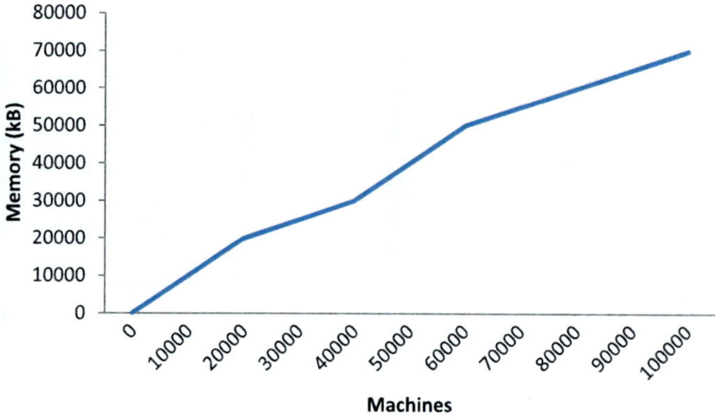

Fig. 16.5 Memory usage in resources instantiation

Nevertheless, in the time-shared situation, the running time of each work changed as the amount of work completed increased. Because the processor core is continuously perspective flipped among the list of planned jobs, this strategy has a considerable impact on completion time. Because the servers were not overburdened at the start of the performance, the first group of 50 jobs was capable of completing faster than the others. Finally, as more jobs were completed, a greater number of hosts became accessible for distribution. As a result, they saw faster reaction times for the activities, as shown in Fig. 16.6.

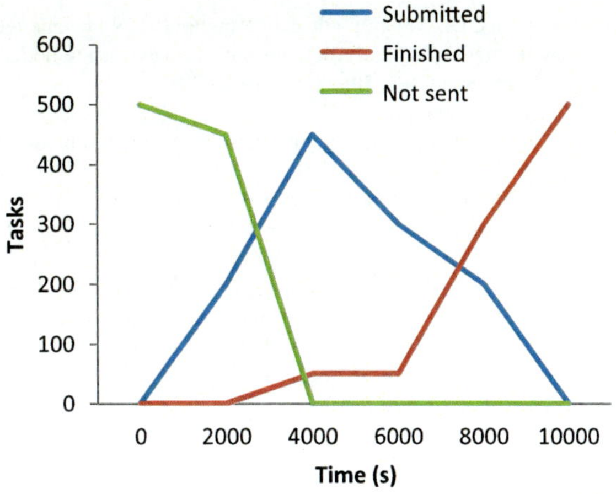

Fig. 16.6 Task execution with time-shared scheduling of tasks

Conclusion

Recent attempts to develop and construct Cloud technologies have centered on establishing innovative methodologies, strategies, including procedures for handling Cloud technologies effectively. To put these recently founded methodologies and regulations to check, researchers require skills that enable them to examine the assumption before deploying it in a controlled setting. Simulation-based initiatives bring numerous advantages, particularly in the case of Cloud technology, where connectivity to connectivity charges a fee in actual money because they enable Cloud programmers to develop the effectiveness of their configuration management and delivery of services policy initiatives in a repetitive and easy to control surroundings at no expense, and to optimize performance bottlenecks before actually launching on real Clouds. To that purpose, they created the CloudSim technology, a modeling and modeling platform for next-generation Clouds. This is ideal as a research instrument that could manage the complications coming from simulations since it is a configurable instrument that enables the expansion and formulation of rules in all elements of the software platform. They want to add additional billing and supply rules to CloudSim in the coming to provide built-in support for simulating the currently existing Clouds.

References

1. Serafin, F., David, O., Carlson, J.R., Green, T.R., Rigon, R.: Bridging technology transfer boundaries: Integrated cloud services deliver results of nonlinear process models as surrogate model ensembles. Environ. Model. Softw. **146**, 105231 (2021)
2. Mangas, A.G., Alonso, F.J.S., Martínez, D.F.G., Díaz, F.D.: WoTemu: an emulation framework for edge computing architectures based on the Web of Things. Comput. Netw. 108868 (2022)

3. Mourtzis, D., Angelopoulos, J., Panopoulos, N.: Challenges and opportunities for integrating augmented reality and computational fluid dynamics modeling under the framework of industry 4.0. Procedia CIRP **106**, 215–220 (2022)
4. Zheng, W., Muthu, B., Kadry, S.N.: Research on the design of analytical communication and information model for teaching resources with the cloud-sharing platform. Comput. Appl. Eng. Educ. **29**(2), 359–369 (2021)
5. Liu, D., Chang, Q.: Research on integration framework and service adaptation technology of space system simulation test-bed. In: Journal of Physics: Conference Series, vol. 2133, no. 1, p. 012038. IOP Publishing, Nov 2021
6. Xiong, L., Li, M.: Behavioral modeling based on cloud computing and target user recommendation for English cloud classroom. Microprocess. Microsyst. **80**, 103587 (2021)
7. Shekhar, C.A., Sharvani, G.S.: MTLBP: a novel framework to assess multi-tenant load balance in cloud computing for cost-effective resource allocation. Wireless Pers. Commun. **120**(2), 1873–1893 (2021)
8. Bhasha, A.C., Balamurugan, K.: Studies on mechanical properties of Al6061/RHC/TiC hybrid composite. Int. J. Lightweight Mater. Manuf. **4**(4), 405–415 (2021)
9. Lee, K., Shin, J., Kwon, S., Cho, C.S., Chung, S.: BIM environment based virtual desktop infrastructure (VDI) resource optimization system for small to medium-sized architectural design firms. Appl. Sci. **11**(13), 6160 (2021)
10. Lindsay, D., Yeung, G., Elkhatib, Y., Garraghan, P.: An empirical study of inter-cluster resource orchestration within federated cloud clusters. In: 2021 IEEE International Conference on Joint Cloud Computing (JCC), pp. 44–50. IEEE, Aug 2021
11. Balamurugan, K., Uthayakumar, M., Sankar, S., Hareesh, U.S., Warrier, K.G.K.: Preparation, characterization, and machining of LaPO4-Y2O3 composite by abrasive water jet machine. Int. J. Comput. Aided Eng. Technol. **10**(6), 684–697 (2018)
12. Deepthi, T., Balamurugan, K., Balamurugan, P.: Parametric studies of abrasive waterjet machining parameters on Al/LaPO4 using response surface method. In: IOP Conference Series: Materials Science and Engineering, vol. 988, no. 1, p. 012018. IOP Publishing (2020)
13. Karnakanti, S., Radhika, K.: Report generation mechanism-based infrastructure as a service (IAAS) framework designing-review (2021)
14. Das, A., Gupta, J.T., Reddy, N.S., Hallymysore, P.: Optimizing constraint-driven container orchestration policies in a cloud environment. In: 2022 IEEE 12th Annual Computing and Communication Workshop and Conference (CCWC), pp. 0552–0560. IEEE, Jan 2022
15. Mazidi, A., Mahdavi, M., Roshanfar, F.: An autonomic decision tree-based and deadline-constraint resource provisioning in cloud applications. Concurrency Comput.: Practice Experience **33**(10), e6196 (2021)

Chapter 17
Data Mining Approach for Prediction of Various Risk Factors in Supply Chain Management

D. R. Kumar Raja, G. Hemanth Kumar, and P. Lakshmi Sagar

Introduction

As a consequence of enhanced internationalization, customer requests, and brief life cycle in SC systems, the incidence, intensity, and variation of project risks were rising [1]. Suppliers were vulnerable to a variety of internal and external hazards, each with its own set of characteristics, chances, and consequences. Uncertainty needs, delivery problems, unpredictable currency exchange, political unrest, changing customer marketplaces, and even unforeseen incidents like work accidents, cyber-attacks, natural calamities, and terrorism could all contribute to all of these hazards [2–4]. Due to the complexities of these systems, measuring the trigger elements of SC hazards, their linkages, and repercussions was extremely difficult. Advancements in information and communication technology are continually expanding the scope, diversity, quantity, and quantity of information.

Organizations were increasingly using information judgment processes as a consequence of technical advancements and enhanced information congestion. To make more efficient, accurate, and decisions at the right time, an increasing number of firms were turning to Data Analytics and Data Mining methodologies [5]. DM was essential for learning about possible SC risk variables, their origins, consequences,

D. R. Kumar Raja (✉)
Associate Professor, School of Computer Science and Engineering, REVA University, Bengaluru, India
e-mail: kumarraja.raj@gmail.com

G. Hemanth Kumar
Assistant Professor, School of Electronics and Communication Engineering, REVA University, Bengaluru, India

P. Lakshmi Sagar
Assistant Professor, Department of CSSE, Sree Vidyanikethan Engineering College, Tirupati, India

Ch. Satyanarayana et al. (eds.), *Proceedings of the International Conference on Computer Vision, High Performance Computing, Smart Devices and Networks*, Advanced Technologies and Societal Change,
https://doi.org/10.1007/978-981-19-4044-6_18

and interconnections [6]. To construct reactive and proactive systems, DM methods can be implemented in different phases of SCRM. DM approaches have been used in a range of disciplines to discover and grade based occurring in different research disciplines. Despite increased emphasis on research, scientific literature was limited and widely disseminated [7].

For SCRM, there seems to be a lack of a comprehensive structure for using large amounts of data. The conceptual methodology was based on a survey of available literature and also the expert judgment of SC executives and BI specialists [8].

Related Works

This methodical developed a comprehensive recommendation by trying to cover numerous operations of information risk assessment, like as problem that may occur identifier, hazard information gathering, constructing a hazard database system, risk assessment and management group characteristics, converting of the SCRM issue into a DM issue, and risk assessment interpretation of results of DM methodologies [9]. A research analysis in the large equipment industry was used to verify the framework, plus crucial recommendations for implementing the suggested strategy were made [10]. The methodology established was expected to serve as a baseline for future scientific work on the use of DM methods in SCRM.

The goal of SCRM would be to gain a better understanding of hazards and their consequences and also to implement preventative health measures to mitigate them. Today's information and communications technology enable the collection, storage, and analysis of a wide range of dangerous information from a multitude of locations. As a consequence of these advances, the idea of "Risk Intelligence" has developed [11]. Risks intelligence involves the capability of an organization to recognize, evaluate, appraise, and forecast dangers based on the aforementioned historical facts and expertise. Only a few academic research has looked at problem behaviors or cognition from an SC perspective [12].

Several firms have begun to utilize automation, risk management frameworks to operate in today's experience and understand the economic landscape because it would be evident that they cannot control their hazard without controlling their information and data. Considering the high level of interest in business, academics seem to have been slow to use BI for the control of SC hazards [13]. For strategic planning, BI includes a database, techniques, procedures, and technology for transforming raw numbers into meaningful data and sensitive data. Some of the important equipment from BI [14] include extraction, converts, and loads, data storage, Online Analytical Processing, data analysis, forecasting, and visualization. DM, therefore, plays an important role in creating a BI ecosystem and also has a multitude of uses in various fields.

In the finance and insurance industries, DM approaches were commonly utilized in customer engagement, customer defection predictions, and customer risk assessment [15]. By evaluating customer information, including transaction histories, credit card information, satisfaction levels, and credit quality ratings, DM systems also were employed to continue providing valuable insights into problematic consumers. Present a credit or debit card data-based DM-based personal bankruptcy predictive model [16]. Researchers forecast sequences using pattern mined approaches and would then present a unique model-based k-means grouping strategy to find sequence. By analyzing terms in internet advertising, a model that can predict customer attrition in the smartphone business was recently created. Economic risk management would be another frequent DM application.

Proposed Technique

The goal of this study would be to create a DM-based SCRM system that integrates numerous activities like as purpose of assessment identifier, data analysis, processing and analysis, risk assessment issue transcription into a DM issue, analysis of data utilizing DM methodologies, and interpretation of the findings to recognize smart risk management plan. Understanding in many various disciplines, and also a multi-data collecting and research orientation, were required to create this system. To construct a conceptual framework, the SCRM, DM, document management, and information systems works of literature had first been examined. In constructing such a complicated system, identifying a group of experts and selecting a DM technique were crucial. Primary information from conversations and interactions involving SC and IT specialists were used to develop and verify the conceptual framework. Key elements from DM, document management, and risk assessment were rigorously combined to establish a holistic information SCRM architecture. Figure 17.1 shows the evolution of the DM-based SCRM architecture in a step-by-step manner. The suggested model's major phases seem to be: (i) problem that may occur, identifying, (ii) advancement of a hazard database system to collect and store risk information, and (iii) integration of a DM subsystem, which contains converting the hazard management issue into a DM issue and interpreting the results obtained for risk assessment behavior recognizing and measuring threats and risks necessitates a basic understanding of the SC program's architecture, and also its architectural, economic, and flow of information. As a result, the focal business chooses and applied the theoretical foundation generated from the research. This stage informs us about the company's location within the system, its relationships with other players, the corporation's perceived risk and resiliency measures, the industry's appetite for risk, and proper risk management methods. The 4 sub-tasks in this step have been just as follows:

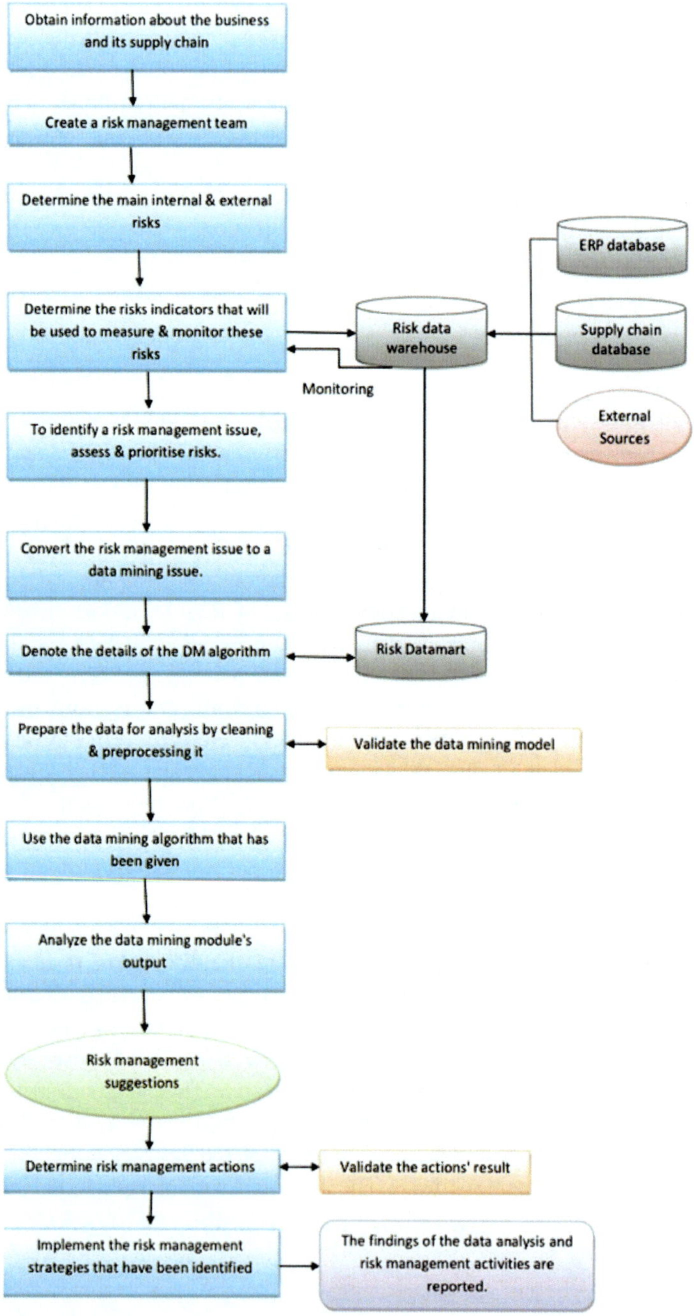

Fig. 17.1 DM-based SCRM model

Results and Discussions

The DM component's goal is to turn vital information into hazard data so that SCRM can make educated judgments. Hazard information was required unless and until it is processed. The DM devices collect hazard information from the system and create a data mart from the RDW for the present DM assessment. The information mart that would be susceptible to the DM algorithms was referred to as a hazard database table in the suggested DM-based SCRM architecture. The hazard data mart seems to be a component of the RDW that would be built by processing, summarizing, and structuring the information according to the DM platform's requirement. The DM-based infrastructure that would be utilized to translate risk data can be seen in Fig. 17.2. Categorization, predictions, analysis, association rule, grouping, outlier detection, and other predictive and descriptive DM activities could be used to analyze risk information. Because every DM job has its capability, the choice that should be made depends on the intended outcome and also the method's requirements. The following difficulties could be addressed using DM tools and methods: detecting issues, predicting hazards and their repercussions, identifying correlations among hazards and some other variables, identifying hazard root causes, clustering problematic objects, etc. Health issues and long-term risk data patterns could be predicted using forecasting algorithms.

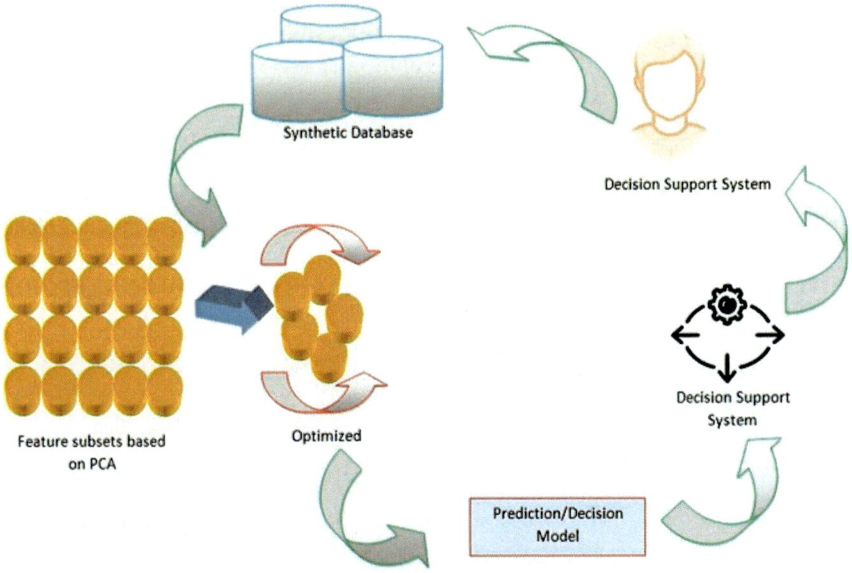

Fig. 17.2 A general DM-based architecture

Interpreting Results

When it comes to evaluating outcomes, visualization tools can be really useful. The researchers' findings could be saved with the hazard information will be stored. The organizational risk management team needs to identify activities for mitigating specific lifestyle factors following identifying and assessing risk information. There seem to be a variety of SCRM methods and tactics to choose from. Subject specialists must assess various supply chain solutions, taking into account costs involved and advantages.

Implementing and Testing

To evaluate risk areas and identify the dispersion of hazards across the organization and also its SC system, personnel from multiple departments were given a hazard appraisal form to fill out. On a ten-point Likert scale, respondents were asked to indicate the likelihood, repercussions, and feasibility of identification of potential threats confronting their firm. Depending on respondent replies, Fig. 17.3 depicts the risk occupational exposure of several operational functions. The most dangerous activities of the organization, as shown in the diagram, were procuring, marketing, and product/project development. Figure 17.4 also shows the top 15 dangers identified in the current business based on the risk prioritization ratings. This research assisted in identifying the dangerous areas of the SC network and provided empirical and Insights for selecting the risk problem for DM analysis.

Numerous flaws in the planned DM-based SCRM system were discovered during implementation. The datasets seem to be the input to the DM methodologies. In conclusion, two major problems have a considerable impact on the validity of the DM component's outcomes: (i) The system must contain comprehensive, precise, clear, relevant, and up-to-date information; (ii) the appropriate metrics for measuring and tracking specified risk variables must be defined. Moreover, the DM-based SCRM

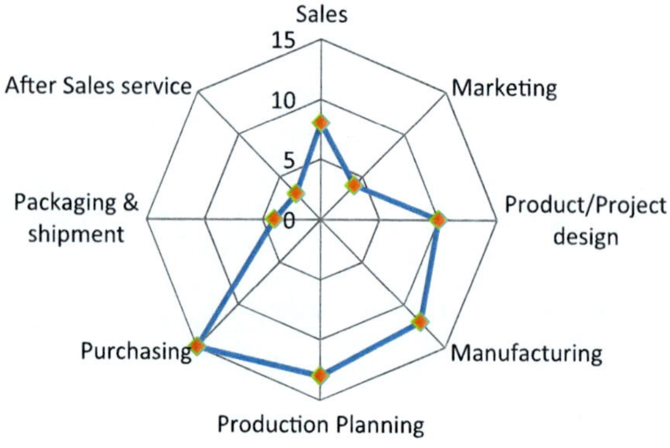

Fig. 17.3 Spider-web diagram of the risk exposure levels

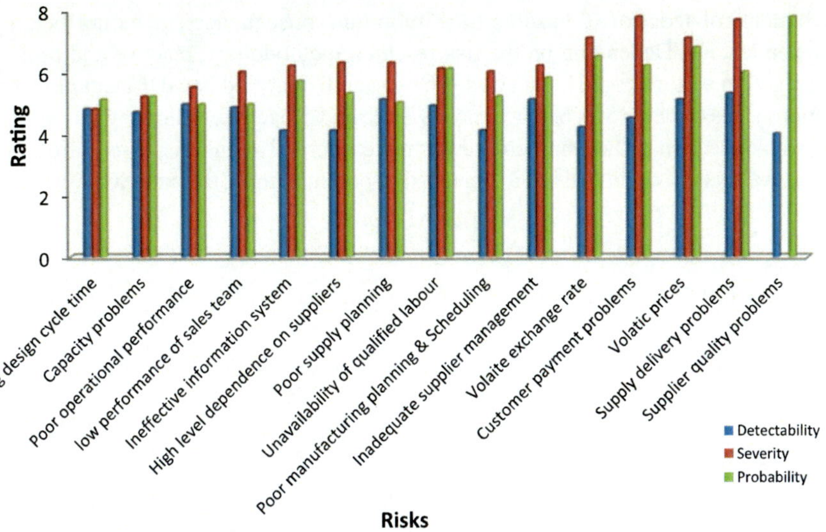

Fig. 17.4 Ratings of the top 15 risks

model's adaptability is contingent on the association's operational technological sophistication.

Conclusion

With advancements in information and communications technology, greater stakeholder involvement, and digitalization of supply chain operations through the adoption of Industry 4.0 technologies, the amounts of information produced and communicated over SC systems seem to be on the increase. SC executives were already being pushed to make informed decisions as a result of these technical advancements and commercial developments. Despite the increasing business interest, academics seem to have been slow to adopt an informed approach for managing SC hazards. The SCRM architecture machine learning-based was unique and has never been handled before. The research presents a holistic method that's also realistic and straightforward to adopt, as proven by a real case instance, by combining multiple system components. A worldwide corporation has partly deployed and evaluated this comprehensive and organized information process. The results and findings of the research study present realistic operational problems and recommendations. The application's application, lessons learned, and constraints were assessed using a single example firm, making it impossible to generalize the framework's applicability, knowledge gained, and limits. The suggested DM-based risk assessment methodology, on the other hand, was supposed to make businesses wiser by offering a better knowledge of SC hazards, their important consequences, including their interdependencies. Depending on the risk research methodology, reactive and proactive mitigation actions could be devised. The research analyzed how DM techniques and approaches can help SCs become more visible and proactive. Following this research, it's also clear that DM can be used to create interactive and dispersed information retrieval systems, and that has a great deal of potential in SCRM and SCM in general.

References

1. Kosasih, E.E., Brintrup, A.: A machine learning approach for predicting hidden links in the supply chain with graph neural networks. Int. J. Prod. Res. 1–14 (2021)
2. Liou, J.J., Chang, M.H., Lo, H.W., Hsu, M.H.: Application of an MCDM model with data mining techniques for green supplier evaluation and selection. Appl. Soft Comput. **109**, 107534 (2021)
3. Kong, J., Yang, C., Wang, J., Wang, X., Zuo, M., Jin, X., Lin, S.: Deep-stacking network approach by multisource data mining for hazardous risk identification in IoT-based intelligent food management systems. Comput. Intell. Neurosci. (2021)
4. Belhadi, A., Kamble, S.S., Mani, V., Benkhati, I., Touriki, F.E.: An ensemble machine learning approach for forecasting credit risk of agricultural SMEs' investments in agriculture 4.0 through supply chain finance. Ann. Oper. Res. 1–29 (2021)
5. Shah, S.M., Lütjen, M., Freitag, M.: Text mining for supply chain risk management in the apparel industry. Appl. Sci. **11**(5), 2323 (2021)
6. Qu, Q., Liu, C., Bao, X.: E-commerce enterprise supply chain financing risk assessment based on linked data mining and edge computing. Mobile Inf. Syst. (2021)
7. Jia, X., Zhang, D.: Prediction of maritime logistics service risks applying soft set-based association rule: an early warning model. Reliab. Eng. Syst. Saf. **207**, 107339 (2021)
8. Liu, Y., Zhang, S., Chen, M., Wu, Y., Chen, Z.: The sustainable development of financial topic detection and trend prediction by data mining. Sustainability **13**(14), 7585 (2021)
9. Tirkolaee, E.B., Sadeghi, S., Mooseloo, F.M., Vandchali, H.R., Aeini, S.: Application of machine learning in supply chain management: a comprehensive overview of the main areas. Math. Probl. Eng. (2021)
10. Garikapati, P., Balamurugan, K., Latchoumi, T.P., Malkapuram, R.: A cluster-profile comparative study on machining AlSi7/63% of SiC hybrid composite using agglomerative hierarchical clustering and k-means. SILICON **13**(4), 961–972 (2021)
11. Latchoumi, T.P., Parthiban, L.: Quasi oppositional dragonfly algorithm for load balancing in cloud computing environment. Wirel. Pers. Commun. 1–18 (2021)
12. Arunkarthikeyan, K., Balamurugan, K.: Performance improvement of cryo treated insert on turning studies of AISI 1018 steel using multi-objective optimization. In: 2020 International Conference on Computational Intelligence for Smart Power System and Sustainable Energy (CISPSSE). IEEE, July 2020, pp. 1–4 (2020)
13. Alipour-Vaezi, M., Aghsami, A., Rabbani, M.: Introducing a novel revenue-sharing contract in media supply chain management using data mining and multi-criteria decision-making methods. Soft Comput. 1–18 (2022)
14. Chinnamahammad Bhasha, A., Balamurugan, K.: Fabrication and property evaluation of Al 6061+x% (RHA+TiC) hybrid metal matrix composite. SN Appl. Sci. **1**(9), 1–9 (2019)
15. Bakhtadze, N., Suleykin, A.: Industrial digital ecosystems: predictive models and architecture development issues. Ann. Rev. Control. **51**, 56–64 (2021)
16. Cheng, K.C., Huang, M.J., Fu, C.K., Wang, K.H., Wang, H.M., Lin, L.H.: Establishing a multiple-criteria decision-making model for stock investment decisions using data mining techniques. Sustainability **13**(6), 3100 (2021)

Chapter 18
Sensor-Based Radio Frequency Identification Technique for the Effective Design Process of Diverse Applications Using WSN

K. Srinivasa Babu, K. Jaya Sri, and Cholleti Jyothi

Introduction

Electronics are now becoming compact and much more sophisticated as segments and sub-networks and communication capabilities advance [1]. As a result of this advancement, ubiquitous computing will become more common in people's daily lives. Radiofrequency identification (RFID) is a vital technology for ubiquitous computing because it allows an entity to be detected from a small distance without the need for user involvement [2]. Different kinds of RFID tags, and communication systems, have recently been accessible incorporate markets around the globe [3, 4].

Even though checking the external conditions is critical in the administration of environment-sensitive goods, current RFID infrastructures & applications lacked this feature [5]. Let's look at a wealth management system that uses RFID as an instance. It is feasible to maintain track of the present position of a specific object using RFID alone [6]. Environmental data, such as humidity levels, is not accessible, though. Wireless sensor nodes, that are made up of numerous small-size computing & sensor systems plus modern communications capabilities, can be used to educate about the

K. Srinivasa Babu (✉)
Professor CSE, Nalla Narasimha Reddy Education Society's Group of Institutions, Hyderabad, Telangana, India
e-mail: Kasturisrinu.babu@gmail.com

K. Jaya Sri
Associate Professor CSE, Nalla Narasimha Reddy Education Society's Group of Institutions, Hyderabad, Telangana, India
e-mail: jayasri.kuchimanchi@gmail.com

C. Jyothi
Assistant Professor CSE, Nalla Narasimha Reddy Education Society's Group of Institutions, Hyderabad, Telangana, India
e-mail: jyothicholleti11@gmail.com

Ch. Satyanarayana et al. (eds.), *Proceedings of the International Conference on Computer Vision, High Performance Computing, Smart Devices and Networks*, Advanced Technologies and Societal Change,
https://doi.org/10.1007/978-981-19-4044-6_19

mitigating factors in such instances [7]. WSNs can gather, combine, and analyze environmental parameters, allowing them to be used in a variety of applications including fire diagnosis and prevention [8]. As a result, by combining RFID devices with WSNs, we may create an object processing and monitoring solution that supports more detailed info on product's surroundings and also their whereabouts.

A few recent studies have looked into ways to connect RFID systems with WSNs. Zhang & colleagues developed systems topologies in which a wireless sensor network's gateways, and sensors, were combined with an RFID reader and tag, correspondingly [9]. Even though the designs could physically merge RFID & sensor systems, the study does not address any engineering problems, such as how to dynamically build an application that takes advantage of both networks' distinct properties [10].

Related Works

Inventory tracking master-planned using RFID & sensor systems [11]. Whenever we considered the restricted detecting range & energy requirements of detectors, nevertheless, combining RFID readers with detectors is not an expense design [12, 13]. In the interconnected system, they examined mobility, therefore, identified three service types: house care, personalized service, & objective service. The results demonstrate how well the networks may be linked to the telecoms infrastructures for these applications. Researchers did not, nevertheless, demonstrate how and where to interpret sensor input and tag identifiers comprehensively and proactively.

To their understanding, no specific consideration of the relationship between an item and its ambient factors has been implemented in the design of a proposed system. In this paper, we offer SARIF, a new framework that allows applications to be created flexibly while masking the intricacies of RFID systems & WSNs. That implies that a user simply has to declare the specifications of the items in the associated applications, & SARIF would automatically handle the environment-sensitive items. The following is how the rest of the article is structured. To create a process of combining, we must first determine the most important needs. Then we go over functional components in SARIF and show how they're implemented. Lastly, they review our findings and make recommendations for further research.

Proposed Methods

In RFID networking and WSNs, contrast client–server networks, where a massive data flow is delivered through servers to users, the major data movement is from smartphones and tablets to a few servers. Devices would have to be able to identify occurrences, and RFID readers, would have to be capable of recognizing tagging before the data is sent to one or more databases. Because when data is collected, a website must integrate it with data from RFID systems and WSNs promptly so that corrective steps may be done. The connectivity mechanism must meet the requirements.

In most WSNs, replacing the charges of the instruments in the fields is not cost-effective, consequently decreasing the power usage of the detectors is a crucial concern. Furthermore, regulating the power consumption across detectors is critical for extending the total lifetime of a WSN. Devices close to doorways must usually send sensory information more often than those further distant. As a result, detectors close to the doorway typically run out of battery more rapidly. If all of the detectors together around the doorway expire, the doorway remains separated, and the cable network existence is over. As a result, the connection architecture requires a lot of energy effectiveness in terms of lowering power consumption and a fair load distribution across instruments.

Integration Framework

The integration servers are a vital component of SARIF that controls all tasks in RFID and WSN connections shown in Fig. 19.1. WSNs are made up of gateway nodes and a sensor network, while RFID systems have an EPC knowledge server, RFID readers, plus RFID tags. An RFID reader detects a tag and sends the tag identification to the EPC information server in RFID networks. The EPC identification servers keep track of each tag identifier's properties and send them to the organization's network. This integration server starts a certain job in the WSN based on the data collected from the RFID networks. This same integration server accomplishes this by retrieving significant aspects and constructing a signal, that is then dispersed to the WSN via one or more entry points. Following then, each instrument designated to the job begins monitoring ambient variables and analyzing them using incident data mining algorithms. The sensors send raw sensual information to the server based on the procedure indicated by the assignment.

Implementation

Researchers created a SARIF sample testbed with 16 MICA2 motes and two portals for measuring investigations, as shown in Fig. 19.2. A sensors boards featuring a range of primary objectives, including lighting, warmth, & audio, are made up of an MTS 300 and a MICA2 mote. The detector operates TinyOS, which has been tweaked for SARIF's sensor systems integration. The RFID networks are simulated using DRIVE. IBM's DRIVE is a cross-platform framework that makes developing RFID solutions easier by allowing the creation of model-based parts. Leading to a shortage of space, they have left out the intricacies of DRIVE. Rather, we'll go through the WSN and the integrating servers in further detail.

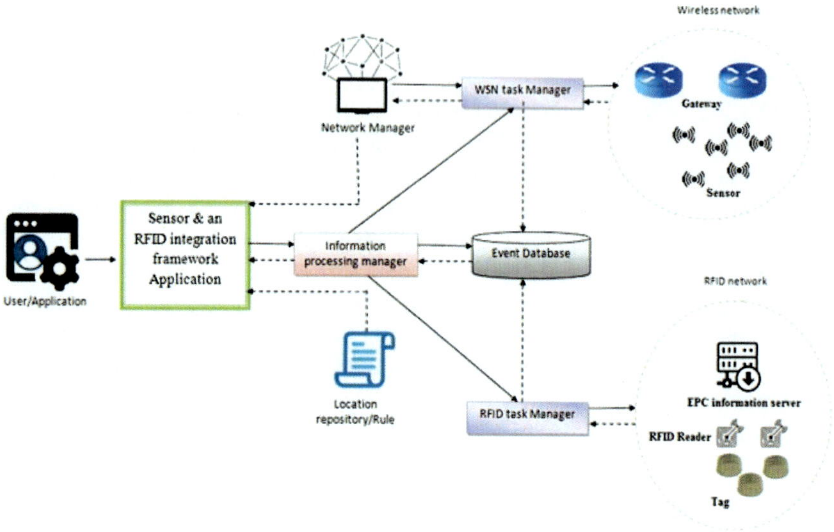

Fig. 19.1 Functional components of SARIF

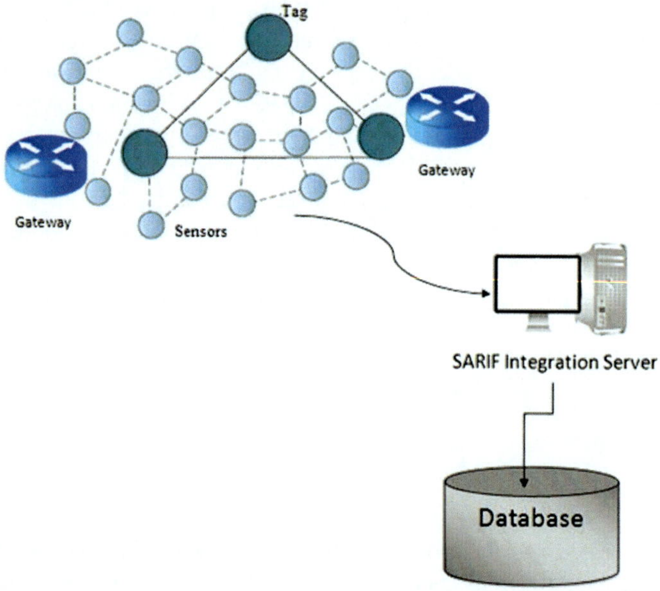

Fig. 19.2 SARIF prototype architecture

Wireless Sensor Network Protocols

Researchers design a task assignment method and a dynamic load balancing strategy for effective implementation, that can improve the battery performance of the detectors while also ensuring that activities are delivered accurately. Project assignments, administration confirmation, and completion are the 3 phases of the job scheduling system. This integration server produces a sensor task by parsing instructions from the RFID networks or a request from customers in the first stage. The WSN task scheduler then delivers a sensory subtask to the access points, that convert the work package into communication that may be sent to the necessary sensor network. Each detector shall evaluate whether the job belongs to the detector in the second stage, after receiving the job. If a genuine assignment is completed, the sensor will send an acceptance communication to the gateways, that is sent to the WSN device manager. If a duplicated job is received, it should be eliminated. Based on application objectives, if the integration servers do not get an acceptance response from certain of the devices, it may retransmission the job to these detectors. The work procedure is ended in the final stage when the management capability expires.

Observations

The incoming traffic percentage for each doorway and the fraction of live nodes for each plan are shown in Figs. 19.3 and 19.4. In the resource scheduling method, the packet delivery proportion is virtually equally divided across the two gates, but in the hop distance-based method, the incoming traffic proportion at doorway A is substantially greater than the one at destination B. This is because, under the load-balancing system, all sensors choose one of the doorways based on the forward's probabilities, which favors a doorway with greater remaining fuel reserves among its one-hop neighbors. In the hop distance-based approach, on the other extreme, detectors automatically select the closest entrance. That load-balancing technique promotes increased energy conservation by dividing the network congestion approximately evenly between the entrances, as shown in Fig. 19.4. Since there are fewer instruments surrounding doorway B, and they are overwhelmed with communication, the detectors close doorway B perish more quickly than those close doorway A in the hopping distance-based method. The volume of traffic is not skewed to a certain doorway in the load balancing system since detectors pick a bridge related to the energy levels of each gateway's one-hop neighbors.

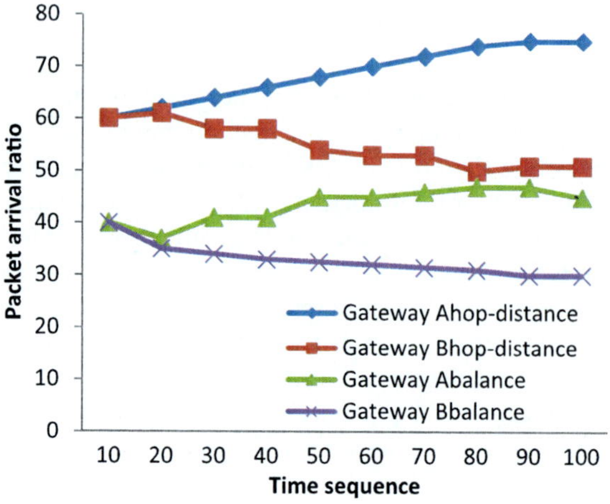

Fig. 19.3 Packets gateway proportion

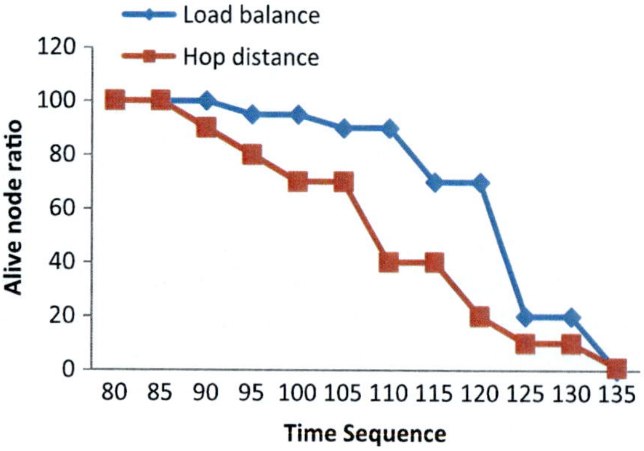

Fig. 19.4 Sensor and alive node ration comparison

Conclusion

Throughout this paper, they offer SARIF, a new framework for monitoring & managing environment-sensitive objects that incorporates RFID systems with WSNs. To combine both systems, researchers initially determined critical needs and outlined the essential features of SARIF. They proved that SARIF could achieve energy consumption through task scheduling by developing the prototype. They plan to expand the experiment in the upcoming by including an actual RFID system.

References

1. Zhu, X.: Complex event detection for commodity distribution internet of things model incorporating radio frequency identification and wireless sensor network. Futur. Gener. Comput. Syst. **125**, 100–111 (2021)
2. Kluczek, A., Gladysz, B., Ejsmont, K.: Application of lifecycle measures for an integrated method of environmental sustainability assessment of radio frequency identification and wireless sensor networks. Energies **14**(10), 2794 (2021)
3. Subasini, C.A., Karuppiah, S.P., Sheeba, A., Padmakala, S.: Developing an attack detection framework for wireless sensor network-based healthcare applications using a hybrid convolutional neural network. Trans. Emerg. Telecommun. Technol. **32**(11), e4336 (2021)
4. Dastres, R., Soori, M., Asamel, M.: Radio Frequency Identification (RFID) based wireless manufacturing systems, a review. Independent J. Managem. Prod. **13**, 258–290 (2022)
5. Mysorewala, M.F., Cheded, L., Aliyu, A.: Review of energy harvesting techniques in wireless sensor-based pipeline monitoring networks. Renew. Sustain. Energy Rev. **157**, 112046 (2022)
6. Rahman, L.F., Alam, L., Marufuzzaman, M., Sumaila, U.R.: Traceability of sustainability and safety in fishery supply chain management systems using radio frequency identification technology. Foods **10**(10), 2265 (2021)
7. Khalifeh, A., Saadeh, M., Darabkh, K.A., Nagaradjane, P.: Radio frequency-based wireless charging for unsupervised clustered WSN: system implementation and experimental evaluation. Energies **14**(7), 1829 (2021)
8. Colaco, J., Lohani, R.: Multi-band microstrip square patch antenna design for IoT-based RFID technology and its various applications. In: 2021 6th International Conference for Convergence in Technology (I2CT), April, pp. 1–7. IEEE (2021)
9. Latchoumi, T.P., Parthiban, L.: Quasi oppositional dragonfly algorithm for load balancing in cloud computing environment. Wireless Personal Commun 1–18 (2021)
10. Muppalaneni, N.B., Prathima, C.: A secure smart shopping cart using RFID tag in IoT. In: Shakya, S., Balas, V.E., Haoxiang, W., Baig, Z. (eds.) Proceedings of International Conference on Sustainable Expert Systems. Lecture Notes in Networks and Systems, vol 176. Springer, Singapore (2021)
11. Chen, Y., Chen, G.: Optimization of the intelligent asset management system based on WSN and RFID technology. J. Sensors (2022)
12. Sivakumar, T.B., Hussain, S.H., Kanmani, A., Babu, M.A.: Surveillance robot for health care applications using IoT and wireless sensor network. In: Materials Today: Proceedings (2021)
13. Amutha, J., Sharma, S., Sharma, S.K.: Strategies based on various aspects of clustering in wireless sensor networks using classical, optimization, and machine learning techniques: review, taxonomy, research findings, challenges, and future directions. Comput. Sci. Rev. **40**, 100376 (2021)

Chapter 19
Reliable Smart Grid Framework Designs Through Data Processing and Analysis Process

Chandra Sekhar Akula, Ch. Prathima, and Asadi Srinivasulu

Introduction

The increasing rise in population and energy use in recent times has enabled changes in the power distribution situation, as well as an increase in energy demand [1–3]. That would be to say, the traditional grid's technology faces several issues. Thankfully, the growth of pervasive computing and mobile internet systems has led to the emergence of a new power distribution architecture known as an advanced grid. Power systems could effectively increase power providing and improving while also addressing current challenges produced by conventional grids [4]. Smart grids could also provide outstanding service and provide connectivity to many sources of power generating [5]. As a result, the smart grid's capacity to self-heal and withstand attacks is critical in promoting a smart grid. Smart grids use a unified two-way or multiple network system to transmit power information, which seems to have specific communication infrastructure needs [6]. However, the multiple communications impact is poor, with many potential flaws; big data method can significantly improve communication precision and effectiveness; therefore, big data technology should indeed be vigorously integrated into smart grid development, the building of a new switching devices telecommunications network, to fix the issues of communication in smart grid [7, 8].

C. S. Akula (✉)
Avanthi Institute of Engineering and Technology, Vizianagaram District, India
e-mail: chanduforu@gmail.com

Ch. Prathima
School of Computing, Mohan Babu University, Tirupathi, India
e-mail: chilukuriprathi@gmail.com

A. Srinivasulu
BlueCrest University, 1000 Monrovia, Liberia

Ch. Satyanarayana et al. (eds.), *Proceedings of the International Conference on Computer Vision, High Performance Computing, Smart Devices and Networks*, Advanced Technologies and Societal Change,
https://doi.org/10.1007/978-981-19-4044-6_20

In the world of big data, data compression technologies are gaining a lot of traction. Depending on the enhanced faulty processing signals, a real-time data reduction and rebuilding technique has been developed [9]. To achieve data compression techniques in energy systems, they used linear integers communication [10].

For smart grid information, a regular 2-D lifting wavelet reduction method is used. The steady flow constants of the power distribution compression algorithms technique [11] are derived from studies. Furthermore, this technique could effectively increase the quality of wireless technology, network connectivity, including storage capacity for determining insulating sub discharges, power dissipation, with a fast sample rate.

In the world of big data, compression algorithms software is gaining a lot of traction. Depending on the enhanced faulty processing signal, a real-time data reduction and rebuilding method. To achieve data compression techniques in the electricity system, researchers used linear numerical communication. For smart grid information, a periodic 2-D lifting wavelet decompression method is used. The quasi-steady characteristics of the electric grid lossless encoding method [12] are based on this work. The technique can effectively enhance the quality of wireless transmission, communication bandwidth, and volume of data for determining insulating sub discharged power losses and fast sampling rate.

Related Works

The employment of a significant range of smart equipment to connect the electrical method to improve knowledgeable, autonomous electricity supply functioning is referred to as smart grid knowledge. Sensors, judgment support systems, sophisticated industrial automation equipment, and inertial measurement units are all fully incorporated into numerous smart network technologies of the smart grid. The intelligence substation, intelligence content delivery system, intelligence scheduling technology, intelligent energy meter, and other components make up the core framework. This smart grid could lower electricity supply operating costs, increase electricity supply stiffness, minimize electricity supply losses, as well as ensure the power base load is economical, secure, and dependable operational. Its smart grid offers different electricity generating accessibility, which makes it easier to utilize renewed energy, and it has a self-healing capability and attacking tolerance, significantly lowering the rate of electric grid failures. Smart grids were originally introduced in 2005, and they have since become a standard for current electric grid development.

Humankind has reached the information age in the twenty-first decade, with information systems being incorporated into numerous disciplines as well as the volumes of information created in group interaction increasing. In the information era, the industrial base is increasingly giving way to the globalized world. In this context, whoever can have the data resources, the marketplace, as well as the benefit of the big data era has arrived, commercial factors, economic interests, and other sectors of all group interaction can no longer function without information.

Proposed Framework

Those aforesaid characteristics and difficulties of smart grid plus big data technologies have inspired me. That part focuses on introducing a unique paradigm for large processing and analyzing in the context of smart grids. Our suggested structure for massive processing and analysis in smart grid is shown in Fig. 19.1. Having created the information collecting as well as storing modules to address the fact that smart grid information cannot be handled in real-time through the cloud service. Many front-end machines are specially installed across from access to data as well as information processing. These front-end systems are also in charge of collecting the warning signals or information that has been discovered. The obtained huge data from the smart grid would next be analyzed. That preprocessing effort includes wavelet transformation and cloud processing. The key concept here is that the SVD processes were hosted online. The coarser smart grid's enormous amount of data would be significantly lowered as well as managed due to the cloud's tremendous processing capabilities. The top layer is the responsibility of evaluating and extracting usable knowledge from the processed data, as well as offering more sophisticated solutions to customers or electrical firms. The abstracted intelligence community, for instance, is being used in generating controllers, energy selling mechanisms, electrical operational structures, situation surveillance cameras, including threat assessment structures.

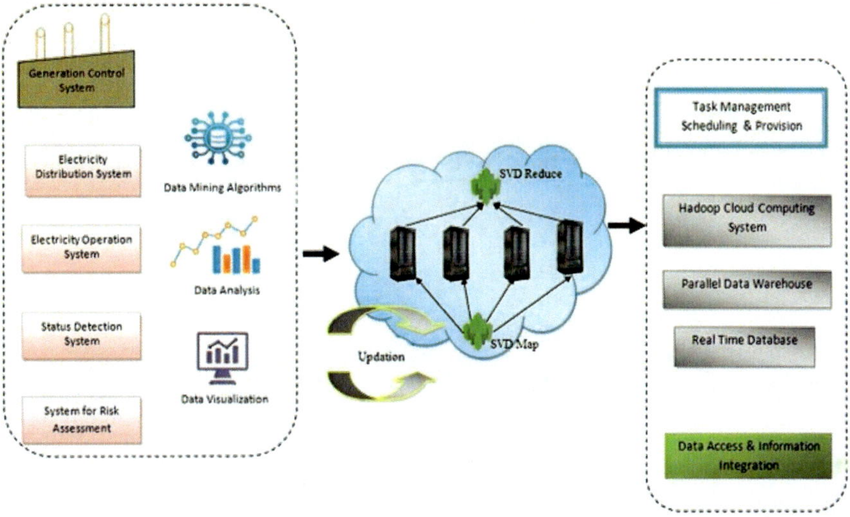

Fig. 19.1 Proposed architecture

Observations

Researchers use load demand forecasting as a study case throughout this part to validate the practicality and efficacy of our suggested system. Because residential energy demand forecasts are a common need for the Power Network Organization, it may give information for decision-making for scheduled centers as well as provide effective power unit recommendations, thereby enhancing the program's safe and secure environment.

Researchers gathered information on energy load utilization for an anonymized consumer in Ahmedabad city (between 3 March as well as 24 March 2020) (see Table 19.1).

They began by collecting information on electrical load utilization for the prospective customer in Ahmedabad City using smart embedded sensors. Then, using our SVD technique, we remove the noise from the signal. As a result, data about routine electrical load utilization is screened out of the program. Afterward when, for projecting future electrical loads, the second exponential smoothing process is used. The projected energy load or pattern may be utilized to aid the electricity supply organization in making the best option for energy programming including choosing the best pricing approach. Using this method, designers could easily purchase the expected conclusions in our research study regarding the pattern of electrical load consumption with forecast outcomes. The segments lines that go within the rectangle in Fig. 19.2 represent the predicted power load in the next 3 time periods (It's worth noting that the time-stamping were set up as per user-defined criteria, including a 3-day delay among authentication information) (Fig. 19.3).

Table 19.1 The electricity load usage

Time	Electricity load
T1	63
T2	63.5
T3	61.4
T4	60.1
T5	60.3
T6	60.9
T7	60.5
T8	61
T9	61
T10	60.7
T11	61.9
T12	63
T13	62.6
T14	65.5
T15	70.2

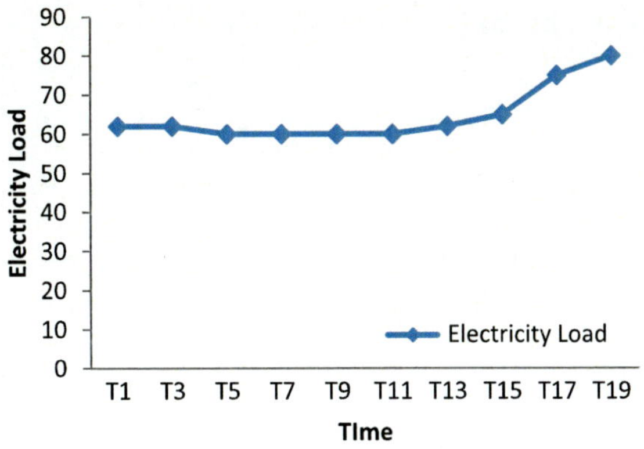

Fig. 19.2 The trend of electricity load usage

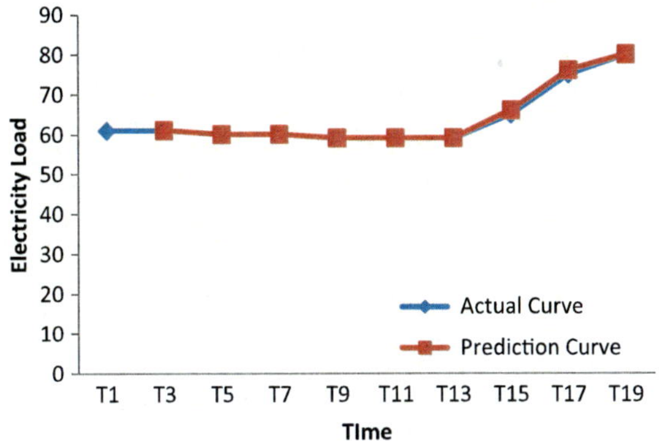

Fig. 19.3 The trend of electricity load usage and prediction

Experimental measurements, researchers may use the increased organizational to forecast the future power demand and to provide the best decision-making approach on both the firm and client ends (Table 19.2).

Because this user's tendency is upward, experts might advise this user to moderate their electricity use in everyday life, as the pricing follows a step price strategy. In terms of balancing the demand as well as ensuring the stability of the system, the power network organization must reassess the programming of neighboring electrical load utilization patterns.

Table 19.2 Result analysis

Time	T17	T18	T19
Electricity load	76.1	77.9	80.8

Conclusions

That research proposed individual places for large processing and analyzing in smart grid, represent the current deficiencies of data transport, storing, and processing according to the properties of smart grid. Having designed an efficient data processing and reception architecture via network connectivity and incorporation, in particular. A prime example of user power demand predictions was undertaken with this proposed methodology to validate its practicality and efficacy.

References

1. Jha, A.V., Appasani, B., Ghazali, A.N., Pattanayak, P., Gurjar, D.S., Kabalci, E., Mohanta, D.K.: Smart grid cyber-physical systems: communication technologies, standards, and challenges. Wireless Netw. **27**(4), 2595–2613 (2021)
2. Chehri, A., Fofana, I., Yang, X.: Security risk modeling in smart grid critical infrastructures in the era of big data and artificial intelligence. Sustainability **13**(6), 3196 (2021)
3. Fan, D., Ren, Y., Feng, Q., Liu, Y., Wang, Z., Lin, J.: Restoration of smart grids: current status, challenges, and opportunities. Renew. Sustain. Energy Rev. **143**, 110909 (2021)
4. Muzumdar, A., Modi, C., Madhu, G.M., Vyjayanthi, C.: A trustworthy and incentivized smart grid energy trading framework using distributed ledger and smart contracts. J. Netw. Comput. Appl. **183**, 103074 (2021)
5. Wang, Z., Jiang, D., Wang, F., Lv, Z., Nowak, R.: A polymorphic heterogeneous security architecture for edge-enabled smart grids. Sustain. Cities Soc. **67**, 102661 (2021)
6. Elsisi, M., Mahmoud, K., Lehtonen, M., Darwish, M.M.: Reliable industry 4.0 based on machine learning and IOT for analyzing, monitoring, and securing smart meters. Sensors **21**(2), 487 (2021)
7. Omitaomu, O.A., Niu, H.: Artificial intelligence techniques in smart grid: a survey. Smart Cities **4**(2), 548–568 (2021)
8. Jamil, F., Iqbal, N., Ahmad, S., Kim, D.: Peer-to-peer energy trading mechanism based on blockchain and machine learning for sustainable electrical power supply in smart grid. IEEE Access **9**, 39193–39217 (2021)
9. Singh, P., Masud, M., Hossain, M.S., Kaur, A.: Blockchain and homomorphic encryption-based privacy-preserving data aggregation model in smart grid. Comput. Electr. Eng. **93**, 107209 (2021)
10. Hashmi, S.A., Ali, C.F., Zafar, S.: Internet of things and cloud computing-based energy management system for demand-side management in smart grid. Int. J. Energy Res. **45**(1), 1007–1022 (2021)
11. Völker, B., Reinhardt, A., Faustine, A., Pereira, L.: Watt's up at home? Smart meter data analytics from a consumer-centric perspective. Energies **14**(3), 719 (2021)
12. Jeyaraj, P.R., Nadar, E.R.S.: Computer-assisted demand-side energy management in residential smart grid employing novel pooling deep learning algorithm. Int. J. Energy Res. **45**(5), 7961–7973 (2021)

Chapter 20
A Novel and Self Adapting Machine Learning Approach of ECG Signal Classification in Association with Cardiac Arrhythmia

P. Rama Santosh Naidu, Golagani Lavanaya Devi, Venkata Ramana Kondapalli, and Ramesh Neelapu

Introduction

Objective

In this project, we classify heart arrhythmias based on the ECG (Electrocardiography) data. We evaluated the proposed method on Physionet's MIT-BIH dataset. MIT-BIH is a widely used dataset for ECG classification. Based on the ECG data, we divided patients into 16 groups. We used many machine learning models to classify the data for this categorization. The major goal of this study is to identify arrhythmias in various individuals and categorize them into appropriate groups so that appropriate treatments may be taken (Table 20.1).

P. Rama Santosh Naidu (✉) · G. Lavanaya Devi · V. R. Kondapalli · R. Neelapu
Department of Computer Science and Systems Engineering, Andhra University College of Engineering (A), Visakhapatnam, Andhra Pradesh, India
e-mail: prsn1988@gmail.com

G. Lavanaya Devi
e-mail: lavanyadevig@yahoo.co.in

V. R. Kondapalli
e-mail: dr.kondapallivr@andhrauniversity.edu.in

R. Neelapu
e-mail: rameshauvsp@gmail.com

Ch. Satyanarayana et al. (eds.), *Proceedings of the International Conference on Computer Vision, High Performance Computing, Smart Devices and Networks*, Advanced Technologies and Societal Change,
https://doi.org/10.1007/978-981-19-4044-6_21

Table 20.1 ECG features

Feature	Description
P wave	The P wave represents depolarization of the atria. Atrial depolarization spreads from the SA node toward the AV node, and from the right atrium to the left atrium
PR interval	The PR interval is measured from the beginning of the P wave to the beginning of the QRS complex. This interval reflects the time the electrical impulse takes to travel from the sinus node through the AV node
QRS complex	The QRS complex represents the rapid depolarization of the right and left ventricles. The ventricles have a large muscle mass compared to the atria, so the QRS complex usually has a much larger amplitude than the P wave
J-point	The J-point is the point at which the QRS complex finishes and the ST segment begins
ST segment	The ST segment connects the QRS complex and the T wave; it represents the period when the ventricles are depolarized
T wave	The T wave represents the repolarization of the ventricles. It is generally upright in all leads except aVR and lead V1
U wave	The U wave is hypothesized to be caused by the repolarization of the interventricular septum. It normally has a low amplitude, and even more often is completely absent

Problem Definition

Patients are divided into 16 groups based on their ECG results. Patient data includes a variety of characteristics such as age, gender, height, and weight. The majority of the patient's information includes results generated from their ECG exam, such as QRS duration.

The different 16 classes are [1, 2]:

CLASS 1. Normal
CLASS 2. Ischemic changes
CLASS 3. Old Anterior Myocardial Infraction
CLASS 4. Old Inferior Myocardial Infraction
CLASS 5. Sinus tachycardia
CLASS 6. Sinus bradycardia
CLASS 7. Ventricular Premature Contraction
CLASS 8. Supraventricular Premature Contraction
CLASS 9. Left bundle branch block
CLASS 10. Right bundle branch block
CLASS 11. Degree Atrioventricular block
CLASS 12. Left ventricular hypertrophy
CLASS 13. Atrial Fibrillation or Flutter
CLASS 14. Others

Existed System

In the existed system doctors examine ECG test reports and have to find out what is type of cardiac problem the patient is facing. It consumes more time and effort to examine the reports.

Proposed System

By anticipating the sort of cardiac arrhythmias, the patient would have, this research aims to automate the work of the doctor. It will take less time, produce reliable results, and allow us to offer the patient the necessary medications.

Literature Survey

This work started mainly due to more heart failures. Nowadays many people are dying due to different types of heart diseases. In this project we classify different arrhythmias, these classes will be useful to detect cardio diseases [1–3].

Cardiac Arrhythmia or heart Arrhythmia is also known as irregular heartbeats [3]. It is a condition where the heartbeat is improper whether it is too slow or too fast. When the electrical impulses in the heart don't operate properly, arrhythmias develop. Arrhythmias can be both dangerous and harmless. The majority of arrhythmias are innocuous and insignificant, but others are exceedingly severe and can lead to death. And many of them which are particularly not dangerous but produce symptoms that can be quite troublesome to your life. A healthy person will never acquire a long-term arrhythmia unless they are exposed to an external trigger or engage in specific activities, such as consuming alcohol or using illegal drugs. If there is an underlying problem, the electrical impulses may not be able to pass through the heart appropriately, raising the risk of arrhythmia. Despite the fact that some people have no symptoms, clinicians can detect arrhythmia during routine ECG testing. Even if a patient detects symptoms, this does not always imply that a major condition exists. Arrhythmias induce four distinct types of symptoms, which are as follows:

- Palpitations
- Dizziness
- Syncope (fainting)
- Cardiac arrest

Sweating, feeling like your heart is racing or fluttering, chest discomfort, feeling like your heartbeat has slowed, and shortness of breath are all symptoms that might occur depending on the kind and degree of your arrhythmia.

The risk factors for Arrhythmia are [4]:

- Old age
- Inherited gene defects
- Heart problems
- Hypothyroidism
- Hypertension
- Obesity
- Illegal Drugs

We need an electrocardiography (ECG) report, as well as the patient's physical exam and medical history, to correctly diagnose cardiac arrhythmia. If an ECG test reveals no arrhythmia, your doctor may do a stress test or a tilt table test.

There are many distinct forms of cardiac rhythm issues, and there are many various therapies for these abnormalities. Even for cardiologists, deciding which medicine to use for a specific arrhythmia can be difficult.

The most common treatment options for cardiac arrhythmias are:

- Anti-Arrhythmia drug therapy
- Pacemakers
- Implantable defibrillators
- Ablation procedures

If deciding on the optimum treatment is challenging, you may be sent to a cardiac electrophysiologist (a cardiologist who specializes in heart rhythm disorders). Even though the majority of arrhythmias generate symptoms, this does not always imply that they are hazardous. Don't worry if you see signs of an arrhythmia; instead, see your doctor. The majority of persons who suffer from arrhythmias have no problems in their daily lives. Many arrhythmias can be treated and prevented by adopting a healthy lifestyle that includes exercise and a nutritious diet. There are many classification algorithms for classifying ECG, EEG signals [5–7].

Proposed Solution

Proposed System

With the help of this proposed system, we try to automate the effort of the physician by predicting the type of cardiac Arrhythmias the patient facing. It will decrease the time and give an accurate result and we can address the patient with the medicines required.

Proposed Algorithm

We utilized the Logistic Regression method to categorize the data. Logistic regression is a technique for describing data and explaining the relationship between one dependent binary variable and one or more independent variables at the nominal, ordinal, interval, or ratio level. A logistic function is used to model a binary dependent variable in a logistic model. A form of predictive analysis is logistic regression. Logistic regression is used to predict the outcome of a categorical dependent variable. As a result, it's necessary to classify the output. Rather of producing precise numbers 0 and 1, the suggested method produces probabilistic values between 0 and 1. To get the highest accuracy, we fit the data to the model and train the data using this method. For data training, we utilize the logistic function, and Logistic Regression may be used to classify observations using various types of data and rapidly find the most effective parameters for classification.

Design and Implementation

The project's system design largely focuses on the data and algorithm training. We immediately train the data and test the fit model using test data. The output will be shown right on the console (Fig. 20.1).

The proposed ECG beat classification approach is evaluated using ECG data from the MIT-BIH arrhythmia data collection (Table 20.2).

Implementation of the Proposed Solution

The implementation contains mainly two parts:

Data Analysis and Data Preprocessing

Data preprocessing is a data mining approach for converting raw data into a format that is both useable and efficient.

Steps involved in data preprocessing

Data cleaning: Data cleaning is the process of eliminating faulty data, structuring the raw data, and filling in the null values before analyzing them.

Removing duplicate values: It will remove all duplicate values and will give a dataset with unique values.

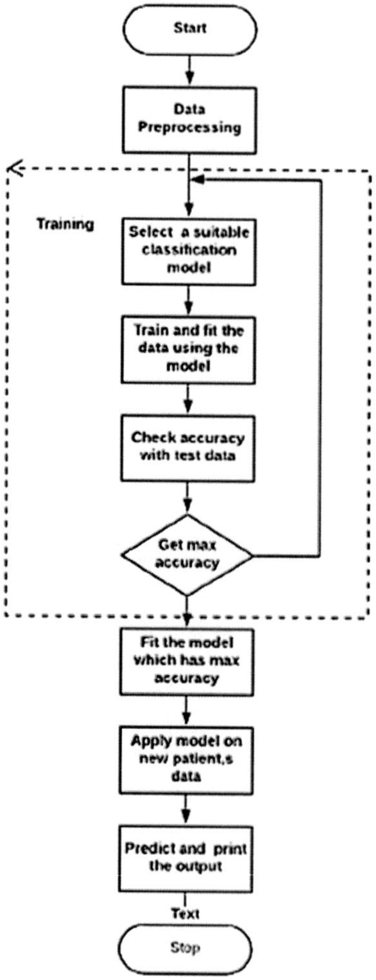

Fig. 20.1 Flow of implementation

Removing null values: To deal with missing values, delete the rows or columns having null values. If more than half of a column's rows are null, the column may be removed.

Feature selection: Feature selection is the process of reducing the number of input variables in a predictive model.

Removing overfitting: When you obtain a decent fit of your model on the training data, but it does not generalize well to fresh, unknown data, you have overfitting.

Table 20.2 ECG class description using AAMI standard

AAMI heartbeat classes	MIT-BIH heartbeat types
Normal (N)	Normal beat Left bundle branch block beat Right bundle branch block beat Atrial escape beat Nodal (junctional) escape beat
Ventricular ectopic (V)	Premature ventricular contraction Ventricular escape beat
Supraventricular ectopic (S)	Atrial premature contraction Aberrated atrial premature beat Nodal (junctional) escape beat Super ventricular premature beat
Fusion (F)	Fusion of non-ectopic and ventricular beat
Unknown (Q)	Paced beat Fusion of paced and normal beat Unclassifiable beat

Training and Fitting a Model

Model Fitting

The classifiers are fitted to the training set. Three different classifiers have been used namely Support Vector Machine, Logistic Regression, *K* Nearest Neighbors. The classification task was divided into 16 groups based on ECG results since each ECG signal type is made of different waves with different amplitudes. Each one of these signals represents an aspect of ECG results. Model is evaluated on the ECG data classifies the data in 16 ECG classes. Train the chosen model with the pre-processed data. More training data increases the accuracy. Training depends on the data and model selected (Fig. 20.2).

Experimental Results and Discussion

ECG data was trained once the Models were created, and the data was separated into training and testing datasets. The training data was used to fit the model, and predictions for the testing data were created. After then, the forecasts were assessed. As we can see in the table, Logistic regression gives more accuracy during the testing and training. As we can see the other two models which we evaluated for the data

```
models = []
models.append(('LR', LogisticRegression()))
models.append(('KNN', KNeighborsClassifier()))
models.append(('SVM', SVC()))

results = []
names = []

for name, model in models:
    kfold = model_selection.KFold(n_splits=10, shuffle=True)
    cv_results = model_selection.cross_val_score(model, X_train, Y_train, cv=kfold, scoring=scoring)
    results.append(cv_results)
    names.append(name)
    msg = "%s: %f (%f)" % (name, cv_results.mean(), cv_results.std())
    print(msg)


LR: 0.623273 (0.080616)
KNN: 0.610060 (0.095494)
SVM: 0.603979 (0.095742)
```

Fig. 20.2 Training output for LR, KNN, and SVM

show lesser accuracy than our proposed algorithm. Both the testing and training accuracies are lesser as compared to the Logistic regression algorithm. Additionally, we also used cross-validation, for this cross-validation step we used 10k-folds cross-validation which divides the data into 10 parts 9 for training and 1 for testing and repeats for all combinations. By performing this cross-validation step, we obtained an error estimate on the test set. After comparing all the accuracies that are obtained from different models, we choose the logistic regression algorithm because it has the highest accuracy in evaluating data and predicting results for the data that is not trained by the model.

Logistic Regression [8, 9]
We utilized the Logistic Regression technique to achieve the categorization. Logistic regression is a technique for describing data and explaining the connection between one dependent binary variable and one or more independent variables at the ordinal, interval, or ratio level. A categorical dependent variable's output is predicted using logistic regression. As a result, the output has to be categorical. The suggested algorithm's output is 0 or 1, however rather than delivering precise values 0 and 1, it delivers probabilistic values between 0 and 1.

Logistic Function (Sigmoid Function) [10, 11]
A mathematical formula for translating expected values into probabilities is the sigmoid function. It transforms any real number between 0 and 1 to another number. The value of the logistic regression must be between 0 and 1, and it must not exceed this limit, else a "S" curve would occur. The S-form curve is the Sigmoid function, often known as the logistic function. In logistic regression, the threshold value is used to indicate the likelihood of either 0 or 1. Values greater than the threshold value are more likely to be 1, whereas those less than the threshold value is more likely to be 0.

Logistic Regression
See Fig. 20.3.

```
LR
0.6703296703296703
['Right bundle branch block ', 'Normal ', 'Ischemic changes ', 'Normal ', 'Normal ', 'Normal ', 'Normal ', 'Sinus bradycardy ', 'Normal ', 'Ischemic changes ', 'Sinus bra
dycardy ', 'Normal ', 'Normal ', 'Normal ', 'Normal ', 'Right bundle branch block ', 'Ischemic changes ', 'Old Anterior Myocardial Infarction ', 'Ischemic changes ', 'Nor
mal ', 'Ischemic changes ', 'Ischemic changes ', 'Normal ', 'Others ', 'Sinus bradycardy ', 'Normal ', 'Normal ', 'Normal ', 'Normal ', 'Normal ', 'Others ', 'Sinus brady
cardy ', 'Right bundle branch block ', 'Normal ', 'Sinus tachycardy ', 'Ischemic changes ', 'Ischemic changes ', 'Ischemic changes ', 'Normal ', 'Ischemic changes ', 'Old
Inferior Myocardial Infarction ', 'Normal ', 'Right bundle branch block ', 'Normal ', 'Normal ', 'Others ', 'Normal ', 'Normal ', 'Normal ', 'Ischemic changes ', 'Normal
', 'Normal ', 'Normal ', 'Right bundle branch block ', 'Normal ', 'Normal ', 'Normal ', 'Normal ', 'Right bundle branch block ', 'Normal ', 'Normal ', 'Right bundle branc
h block ', 'Normal ', 'Right bundle branch block ', 'Normal ', 'Normal ', 'Others ', 'Normal ', 'Normal ', 'Normal ', 'Normal ', 'Normal ', 'Right bundle branch block ',
'Normal ', 'Normal ', 'Normal ', 'Normal ', 'Ischemic changes ', 'Normal ', 'Normal ', 'Normal ', 'Normal ', 'Normal ', 'Others ', 'Ischemic changes ', 'Right bundle bran
ch block ', 'Normal ', 'Sinus tachycardy ', 'Ischemic changes ', 'Normal ', 'Normal ']
[10  1  2  1  1  1  1  6  1  2  6  1  1  1  1 10  2  3  2  1  2  2  1 16
  6  1  1  1  1  1 16  6 10  1  5  2  2  2  1  2  4  1 10  1  1 16  1  1
  1  2  1  1  1 10  1  1  1  1 10  1  1 10  1 10  1  1 16  1  1  1  1  1
 10  1  1  1  1  2  1  1  1  1  1 16  2 10  1  5  2  1  1]
[10  1  2  4  1  1  1  1  1  2  1  1 10  1  1 10  1  3  2  1  5  2  1  9
 10  1  1  1  1  1  1  6 10  1  5  1 15  2  1  2  6  1  1  1  1 16  1  1
  1  1 16  1  1 10  1  6  1  1  5  2  2  1  1 10 14  1  2 14 16  6  1  1
 16  1  1  2  1  2  1  2  1  1  1  9 15 10  1  5  2  1  1]
```

Fig. 20.3 Make predictions on validation dataset (logistic regression)

Confusion Matrix

See Fig. 20.4.

Accuracy of Logistic Regression is 0.6703296703296703.

K-Nearest Neighbor

See Fig. 20.5.

Confusion Matrix

See Fig. 20.6.

Accuracy of *K*-Nearest Neighbor is 0.5934065934065934.

Fig. 20.4 LR evaluation

	precision	recall	f1-score	support
1	0.78	0.84	0.81	50
2	0.57	0.62	0.59	13
3	1.00	1.00	1.00	1
4	0.00	0.00	0.00	1
5	1.00	0.50	0.67	4
6	0.25	0.25	0.25	4
9	0.00	0.00	0.00	2
10	0.60	0.75	0.67	8
14	0.00	0.00	0.00	2
15	0.00	0.00	0.00	2
16	0.20	0.25	0.22	4
accuracy			0.67	91
macro avg	0.40	0.38	0.38	91
weighted avg	0.64	0.67	0.65	91

```
KNN
0.5934065934065934
['Normal ', 'Normal ', 'Normal ', 'Normal ', 'Normal ', 'Normal ', 'Normal ', 'Normal ', 'Normal ', 'Ischemic changes ', 'Normal ', 'Normal ', 'Normal ', 'Normal ', 'Norm
al ', 'Right bundle branch block ', 'Normal ', 'Old Anterior Myocardial Infarction ', 'Normal ', 'Normal ', 'Normal ', 'Normal ', 'Normal ', 'Normal ', 'Normal ', 'Normal
', 'Normal ', 'Normal ', 'Normal ', 'Normal ', 'Normal ', 'Normal ', 'Normal ', 'Normal ', 'Normal ', 'Ischemic changes ', 'Ischemic changes ', 'Ischemic changes ', 'Norm
al ', 'Normal ', 'Normal ', 'Normal ', 'Normal ', 'Normal ', 'Normal ', 'Normal ', 'Normal ', 'Normal ', 'Normal ', 'Normal ', 'Normal ', 'Normal ', 'Normal ', 'Right bun
dle branch block ', 'Normal ', 'Normal ', 'Normal ', 'Normal ', 'Right bundle branch block ', 'Normal ', 'Normal ', 'Normal ', 'Normal ', 'Normal ', 'Normal ', 'Normal ',
'Normal ', 'Normal ', 'Normal ', 'Normal ', 'Normal ', 'Normal ', 'Normal ', 'Normal ', 'Normal ', 'Normal ', 'Normal ', 'Normal ', 'Normal ', 'Normal ', 'Normal ', 'Norm
al ', 'Normal ', 'Ischemic changes ', 'Normal ', 'Normal ', 'Normal ', 'Right bundle branch block ', 'Normal ', 'Normal ', 'Normal ']
[ 1  1  1  1  1  1  1  1  1  2  1  1  1  1  1 10  1  3  1  1  1  1  1  1
  1  1  1  1  1  1  1  1  1  1  1  2  2  2  1  1  1  1  1  1  1  1  1  1
  1  1  1  1  1 10  1  1  1  1 10  1  1  1  1  1  1  1  1  1  1  1  1  1
  1  1  1  1  1  1  1  1  1  1  1  2  1  1  1 10  1  1  1]
[10  1  2  4  1  1  1  1  1  2  1  1 10  1  1 10  1  3  2  1  5  2  1  9
 10  1  1  1  1  1  1  6 10  1  5  1 15  2  1  2  6  1  1  1  1 16  1  1
  1  1 16  1  1 10  1  6  1  1  5  2  2  1  1 10 14  1  2 14 16  6  1  1
 16  1  1  2  1  2  1  2  1  1  1  9 15 10  1  5  2  1  1]
```

Fig. 20.5 Make predictions on validation dataset (*K*-nearest neighbor)

Fig. 20.6 KNN evaluation

	precision	recall	f1-score	support
1	0.60	0.98	0.75	50
2	0.40	0.15	0.22	13
3	1.00	1.00	1.00	1
4	0.00	0.00	0.00	1
5	0.00	0.00	0.00	4
6	0.00	0.00	0.00	4
9	0.00	0.00	0.00	2
10	0.50	0.25	0.33	8
14	0.00	0.00	0.00	2
15	0.00	0.00	0.00	2
16	0.00	0.00	0.00	4
accuracy			0.59	91
macro avg	0.23	0.22	0.21	91
weighted avg	0.44	0.59	0.48	91

SVM

See Fig. 20.7.

Confusion Matrix

See Fig. 20.8.

Accuracy of SVM is 0.6043956043956044.

```
SVM
0.6043956043956044
['Normal ', 'Normal ', 'Normal ', 'Normal ', 'Normal ', 'Normal ', 'Normal ', 'Normal ', 'Normal ', 'Normal ', 'Normal ', 'Normal ', 'Normal ', 'Normal ', 'Normal ', 'Rig
ht bundle branch block ', 'Normal ', 'Old Anterior Myocardial Infarction ', 'Normal ', 'Normal ', 'Normal ', 'Normal ', 'Normal ', 'Normal ', 'Normal ', 'Normal ', 'Norma
l ', 'Normal ', 'Normal ', 'Normal ', 'Normal ', 'Normal ', 'Normal ', 'Normal ', 'Normal ', 'Normal ', 'Normal ', 'Normal ', 'Normal ', 'Normal ', 'Normal ', 'Normal ',
'Normal ', 'Normal ', 'Normal ', 'Normal ', 'Normal ', 'Normal ', 'Normal ', 'Normal ', 'Normal ', 'Normal ', 'Normal ', 'Right bundle branch block ', 'Normal ', 'Normal
', 'Normal ', 'Normal ', 'Normal ', 'Normal ', 'Normal ', 'Normal ', 'Normal ', 'Right bundle branch block ', 'Normal ', 'Normal ', 'Normal ', 'Normal ', 'Normal ', 'Norm
al ', 'Normal ', 'Normal ', 'Normal ', 'Normal ', 'Normal ', 'Normal ', 'Normal ', 'Normal ', 'Normal ', 'Normal ', 'Normal ', 'Normal ', 'Normal ', 'Left bundle branch b
lock ', 'Normal ', 'Normal ', 'Normal ', 'Normal ', 'Normal ', 'Normal ', 'Normal ']
[ 1  1  1  1  1  1  1  1  1  1  1  1  1  1  1 10  1  1  1  1  1  1  1  1
  1  1  1  1  1  1  1  1  1  1  1  1  1  1  1  1  1  1  1  1  1  1  1  1
  1  1  1  1  1 10  1  1  1  1  1  1  1  1  1 10  1  1  1  1  1  1  1  1
  1  1  1  1  1  1  1  1  1  1  1  1  9  1  1  1  1  1  1  1]
[10  1  2  4  1  1  1  1  1  1  2  1  1 10  1  1 10  1  3  2  1  5  2  1  9
 10  1  1  1  1  1  1  6 10  1  5  1 15  2  1  2  6  1  1  1  1 16  1  1
  1  1 16  1  1 10  1  6  1  1  5  2  2  1  1 10 14  1  2 14 16  6  1  1
 16  1  1  2  1  2  1  2  1  1  1  9 15 10  1  5  2  1  1]
```

Fig. 20.7 Make predictions on validation dataset (SVM)

Fig. 20.8 SVM evaluation

	precision	recall	f1-score	support
1	0.58	1.00	0.74	50
2	0.00	0.00	0.00	13
3	1.00	1.00	1.00	1
4	0.00	0.00	0.00	1
5	0.00	0.00	0.00	4
6	0.00	0.00	0.00	4
9	1.00	0.50	0.67	2
10	1.00	0.38	0.55	8
14	0.00	0.00	0.00	2
15	0.00	0.00	0.00	2
16	0.00	0.00	0.00	4
accuracy			0.60	91
macro avg	0.33	0.26	0.27	91
weighted avg	0.44	0.60	0.48	91

Table 20.3 Comparison of accurate prediction of the classifiers

Models	Training accuracy (%)	Testing accuracy (%)
Logistic regression (LR)	63.1	67.03
K-nearest neighbors (KNN)	60.1	59.34
Support vector machine (SVM)	59.5	60.43

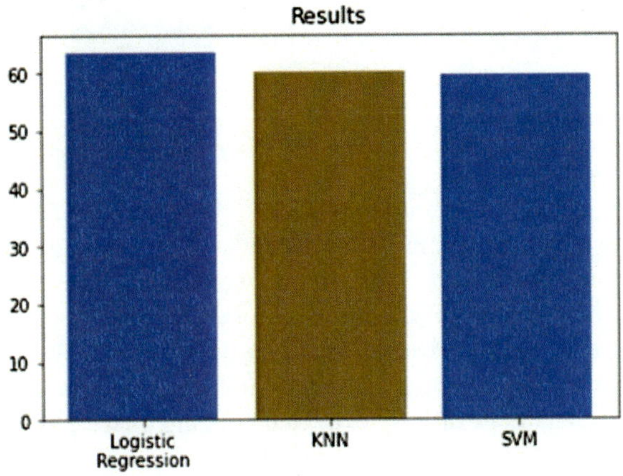

Fig. 20.9 Comparison of accurate prediction of the classifiers (training data)

We evaluated our training data with different models like KNN, SVM, and LR. Among those models, we got the highest accuracy while using Logistic regression so we have chosen it as our model for predicting values.

Table 20.3 shows the performance of each model (Figs. 20.9 and 20.10).

Conclusion and Future Work

In this study, we used machine learning techniques to develop a model for classifying ECG arrhythmias. To be more precise, we've developed a logistic regression model to categorize the various cardiac arrhythmias. The greatest accuracy for both training and testing sets was provided by the logistic regression model, when compared to other models. We can now categorize the arrhythmia issue that a certain patient is experiencing using this model.

It is clear that CLASS 1 (Normal), which comprises the majority of the data, has the greatest impact on the prediction models. Only two or three examples from some of the other classes can be found in the data, making it difficult to understand anything about them and making it impossible to categorize them correctly. As a

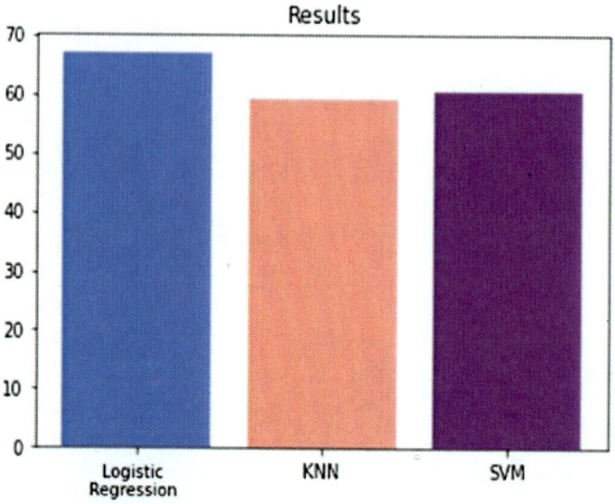

Fig. 20.10 Comparison of accurate prediction of the classifiers (testing data)

result, accumulating more patient instances from different classes is a goal for better future predictions. In the future, we will be able to classify arrhythmia patients more easily and improve the accuracy of all graphs by using smaller data sets. In the future, we may create a user interface for the project that will allow us to easily input patient data using a graphical interface. In future, the data from the patient's ECG report can eventually be directly extracted utilizing image processing and deep learning.

References

1. Rajoub, B.: Machine learning in biomedical signal processing with ECG applications. In: Zgallai, W. (eds.) Developments in Biomedical Engineering and Bioelectronics, Biomedical Signal Processing and Artificial Intelligence in Healthcare, chap. 4, pp. 91–112. Academic Press, Cambridge (2020). https://doi.org/10.1016/B978-0-12-818946-7.00004-4; ISSN 25897527; ISBN 9780128189467
2. Stephenson, R.B.: Electrical activity of the heart. In: Klein, B.G. (eds.) Cunningham's Textbook of Veterinary Physiology, 6th edn., chap. 19, pp. 186–202. W.B. Saunders, Philadelphia (2020). https://doi.org/10.1016/B978-0-323-55227-1.00019-3; ISBN 9780323552271
3. https://www.medicalnewstoday.com/
4. Gatzoulis, M.A., Balaji, S., Webber, S.A., Siu, S.C., Hokanson, J.S., Poile, C., Rosenthal, M., Nakazawa, M., Moller, J.H., Gillette, P.C., Webb, G.D., Redington, A.N.: Risk factors for arrhythmia and sudden cardiac death late after repair of tetralogy of Fallot: a multicentre study. Lancet **356**(9234), 975–981 (2000). https://doi.org/10.1016/S0140-6736(00)02714-8
5. Gurumoorthy, S., Muppalaneni, N.B., Gao, X.Z.: Classification and analysis of EEG using SVM and MRE. In: Computational Intelligence Techniques in Diagnosis of Brain Diseases. Springer Briefs in Applied Sciences and Technology. Springer, Singapore (2018)
6. Gurumoorthy, S., Muppalaneni, N.B., Gao, X.Z.: Analysis of electroencephalogram (EEG) using ANN. In: Computational Intelligence Techniques in Diagnosis of Brain Diseases. Springer Briefs in Applied Sciences and Technology. Springer, Singapore (2018)

7. Gurumoorthy, S., Muppalaneni, N.B., Gao, X.Z.: Analysis of EEG to find Alzheimer's disease using intelligent techniques. In: Computational Intelligence Techniques in Diagnosis of Brain Diseases. Springer Briefs in Applied Sciences and Technology. Springer, Singapore (2018)
8. Alalawi, H.H., Alsuwat, M.S.: Detection of cardiovascular disease using machine learning classification models. IJERT **10**(07) (2021). ISSN 2278-0181
9. Behadada, O., Trovati, M., Chikh, M.A., Bessis, N., Korkontzelos, Y.: Logistic regression multinomial for arrhythmia detection. In: 2016 IEEE 1st International Workshops on Foundations and Applications of Self* Systems (FAS*W), pp. 133–137 (2016). https://doi.org/10.1109/FAS-W.2016.39
10. Hagan, R., Gillan, C.J., Mallett, F.: Comparison of machine learning methods for the classification of cardiovascular disease. Inf. Med. Unlocked **24**, 100606 (2021). https://doi.org/10.1016/j.imu.2021.100606; ISSN 2352-9148
11. Alfaras, M., Soriano, M.C., Ortín, S.: A fast machine learning model for ECG-based heartbeat classification and arrhythmia detection. Front. Phys. **7** (2019). https://doi.org/10.3389/fphy.2019.00103; ISSN 2296-424X